OKOBOJI WETLANDS

A BUR OAK ORIGINAL

OKOBOJI

A LESSON IN
NATURAL HISTORY

UNIVERSITY OF IOWA PRESS IOWA CITY

WETLANDS

by Michael J. Lannoo

With drawings by Danette Pratt and photographs by Carl Kurtz

University of Iowa Press,
Iowa City 52242
Copyright © 1996 by the
University of Iowa Press

Printed in the United States of America

Design by Karen Copp

Printed on acid-free paper

"Frogging in Iowa 1" and "Frogging in
Iowa 2" by William Barrett were originally
published as "Frogging in Iowa" in the
Annals of Iowa 37 (1964): 362–365.
Copyright © 1964 by the State Historical
Society of Iowa. Reprinted by permission
of the publisher.

Jacket/cover: A sora, common but rarely
seen, skitters across an Iowa wetland.
Photo by Carl Kurtz.

Library of Congress
Cataloging-in-Publication Data
Lannoo, Michael J.
Okoboji wetlands: a lesson in natural
  history / by Michael J. Lannoo with
  drawings by Danette Pratt and photo-
  graphs by Carl Kurtz.
p.   cm.—(A Bur Oak original)
Includes bibliographical references
  (p.   ) and index.
ISBN 0-87745-532-5 (cloth),
ISBN 0-87745-533-3 (paper)
1. Wetland ecology—Iowa—Okoboji
  Lakes Region.   2. Natural history—
  Iowa—Okoboji Lakes Region.
  3. Wetlands—Iowa—Okoboji Lakes
  Region—Management.   4. Okoboji
  Lakes Region (Iowa)—History.
  I. Title.   II. Series.
QH105.I8L35   1996
574.5'26325'09777123—dc20   95-43617
                                     CIP

01 00 99 98 97 96 C 5 4 3 2 1
01 00 99 98 97 96 P 5 4 3 2 1

This book is dedicated to the men and women of the Iowa Lakeside Laboratory.

"We sat on a crate of oranges and thought what good [people] most biologists are."

—JOHN STEINBECK, *The Log from the Sea of Cortez*

Proceeds from the sale of this book will be donated to support the student scholarship program at the Iowa Lakeside Laboratory.

CONTENTS

ACKNOWLEDGMENTS

For permission to reprint the articles of Okoboji's past natural historians, I thank Harry Meginnis and Ace Cory, past and present presidents of the Okoboji Protective Association; Paul Rider, editor of the *Journal of the Iowa Academy of Science*; and Marvin Bergman, editor of the *Annals of Iowa*. A special thanks to those who have contributed funds to help cover publication costs: the Okoboji Protective Association's Friends of Lakeside Laboratory (Barb Mendenhall, treasurer); the Dickinson County Soil and Water Conservation Commission through Chris Hoffman and Leo Preston; the Iowa Natural Heritage Foundation through Karmin Blunt; and the Ball State University Faculty Publications Committee. Thanks also to Sue Richter, Harry Meginnis, Phil Brown, and John Synhorst for consultations on funding options.

I cannot praise too highly the activities and foresight of the Okoboji Protective Association (OPA). Again and again in their *Bulletin* (twenty-seven issues published from

1906–1930, and continued for one year in 1940), members question the long-term implications of then present-day environmental activities. Again and again, they increase public awareness of ecological problems and seek solutions. Again and again, they invite Iowa Lakeside Laboratory and Iowa Department of Natural Resources involvement as they work closely with scientists to understand Okoboji's natural history. The only way to measure the environmental changes that have occurred over the course of human settlement is to have a record of this past world. Together with the scientists at Lakeside Lab, the past activists of the OPA have provided us, their descendants, with a snapshot of their world—a picture almost unprecedented in the history of midwestern settlement. In the pages that follow, I hold this snapshot up against our current world for comparison. The OPA remains active today and in fact has enthusiastically and generously supported this book. My colleagues and I look forward to continuing to work with the members of this group in what must be an ongoing effort to monitor and improve Okoboji's environmental quality.

In publications such as this there is usually a disclaimer to the effect that the views of the author do not necessarily reflect the views of the author's institution. Therefore, I should say that the opinions expressed here do not necessarily reflect those of the administration of the Iowa Lakeside Laboratory. I need to point out, however, that the mission of the Iowa Lakeside Laboratory as conceived by its founder, Thomas Macbride, is consistent with a systems approach toward understanding our native natural history. Biodiversity is cherished at Lakeside Lab, and courses that promote an understanding of our natural diversity form the backbone of our curriculum. Programs that restrict or eliminate diversity—such programs are present in Okoboji and are discussed in chapter 5—must be challenged if we are to continue our mission. In this sense, the purpose of this book is consistent with the philosophy of Macbride and his Lakeside Laboratory, this disclaimer notwithstanding.

The notion of wetland ecology that I promote has been refined and polished by conversations with the staff and students at the Iowa Lakeside Laboratory. These lively interactions contribute to the charm of the place and to the enjoyment of our summers. For the privilege of teaching at Lakeside, I thank Richard Bovbjerg and Robert Cruden, our former directors; Arnold van der Valk, our current director; and Bruce Menzel, the chairman of the animal ecology department at Iowa State University. I have not worked alone. Bob Cruden of the botany department at the University of Iowa has contributed the essay on Okoboji insects, especially its dragonflies. Kenneth Lang in the biological sciences department at Humboldt State University in Arcata, California, has written the essays on the seasonality of West Lake Okoboji and on the zooplankton extinction. Ken has also worked with me on the Dickinson County amphibian survey and in obtaining funding support for this project. Richard Baker in the geology department of the University of Iowa has contributed the essays on human impacts and on soil erosion. Dick Bovbjerg has written the essays on mollusk and crayfish biology, and the role that biological field stations play in facilitating the study of animal behavior.

Who can say enough about Carl Kurtz's photographs? I am thrilled to collaborate with Carl on this project and can only hope the quality of the text matches that of his images. Danette Pratt has provided the original illustrations. Danette has that rare combination of a critical technical eye, so necessary for scientific accuracy, and a gifted artistic ability, essential to convey the notion of animals in life. She is also a longtime family friend, and the life she breathes into her drawings is born from her personality. Steve Kennedy, director of the Iowa Great Lakes Maritime Museum, is a valuable Okoboji resource. Captain Steve has kindly allowed me to use a number of his rare photographs documenting historical Okoboji and has provided a tape of an interview with Cap Kennedy, one of the old-time Okoboji fishing guides, but no relation to Steve.

John Maher provided the consultation on federal wetlands regulations. Dick and Debby Baker tracked down articles and references, and introduced me to the work of Gretel Erhlich. Debby Baker provided the historical photographs of Lakeside Lab. Jim Dinsmore sent me the important article by William Barrett. Jane Hey provided the biographical information on T. C. Stephens. Larry Stone, outdoor reporter for the *Des Moines Register*, gave some of my views on amphibian declines their first public airing. Dick Baker, James Baker, Bob Cruden, Lyda Cruden, Robert Dare, Joe Eastman, Evelyn Geiser, Ron Heyer, Ray Johnson, Diane Larson, Bill Leja, Marlyn Miller, John Parsons, Tom Tanner, Sheryl Trinka, and Penelope Wilson have read and critically commented on earlier manuscript drafts. Connie Mutel gave advice on how to write for a nonspecialist audience.

Marilyn Bachmann, formerly at Iowa State University, directed my first amphibian research. Without Dick Bovbjerg, my first mentor, I would not have developed into a professional biologist. (How can you ever express your gratitude to someone for that?) Warm thanks to the many other biologists at Lakeside and elsewhere who have been less involved with this work but who have contributed peripherally. Thanks also to Douglas Triplett and Eugene Wagner of the Muncie Center for Medical Education for allowing me to spend my summers conducting research in Iowa. Mark and Judy Wehrspann, the Lakeside managers, and their family have been supportive and perhaps more tolerant than strictly necessary. Mark was my link to the Lakeside Lab library when I had Okoboji questions but was in Indiana. My parents, Marie and Don, are more responsible for my progress than they know. Finally, I thank Susie, fellow biologist, my wife and partner, and mother to our Pete (may he have the opportunity to see the animals that I have seen, in numbers, where I have seen them). Sue alone, even among my closest friends at Lakeside, knows the thrill and the cost.

OKOBOJI WETLANDS

PREFACE

Our brains emphasize novelty over the familiar, sudden onsets over gradual change. The first northbound robin to arrive in the spring generates so much more attention than the last autumn southbounder to leave; a Thanksgiving snowfall, more excitement (or anxiety, depending on your age and responsibilities) than the final Easter melt. "My how you've grown!"—the doting grandparent's universal exclamation—is simply the result of seeing grandkids infrequently, and therefore comparing their current height with a memory of some relatively distant past. (The extent of this growth is, of course, lost on the constantly vigilant, and ever haggard, parents.)

In the United States, and indeed around the world, humankind's dominion over the earth is resulting in the disappearance of the world's plant and animal species (extinction being the ultimate form of suppression). Scientists have termed this the "biodiversity crisis." The urgency of this crisis has not yet been perceived by the voting public, for various reasons. First, we have crisis overload. When everything today seems to be in crisis, who can become concerned over yet one more? Second, as already noted, our brains are lousy at perceiving incremental change. Nevertheless, an annual five percent decline in an original population over twenty years means extirpation just as surely as does an acute environmental disaster. When local declines are subtle and not perceived, the biodiversity crisis in this age of telecommunications centers on faraway whales, pandas, and rain forests—important, yes, but peripheral to the world one knows.

The way to bring the biodiversity crisis home is to envision our home—our local environment—in the distant past, to visit the world that our grandparents knew when they were children. For most places in the United States, this is a dream because a reliable written record of the local natural history is simply not available. But this is not true everywhere, and it is left to these historical places to remind us of how far we have come, how much we have lost.

Okoboji, in northwestern Iowa, is such a historical place. It may surprise those who know Iowa only from blurred car-window views along Interstate 80, that such a natural region could exist here in this land of nearly endless corn and soybean fields. Yet, as you will see, it is true. Okoboji remains a region of wetlands—of prairie potholes. This is a book that explores what goes on in these wetlands—their biological processes. It highlights wetland animals, especially amphibians, an ecologically important group that is in worldwide decline. This is also a book about changing biodiversity from a historical perspective. In this case, it is about changing biodiversity in a place that we know well. As such, Okoboji becomes a symbol; Okoboji's documented past provides us with a measure of America's lost natural history.

Unfortunately, the loss of biodiversity is not a past problem; it continues today. Therefore, it is not sufficient simply to record historic loss. The causes of these losses must also be identified, and where these processes persist, we should be prepared to reevaluate them with an educated eye toward change. Diminished biodiversity is perceived as one inevitable cost of a changing world driven by the notion of human progress. But is this trade-off necessary? Must these two concepts—natural destruction and human progress—be inexorably linked, or can human progress occur without environmental degradation? Are we, as a people, even environmentally aware enough to make this distinction? In Okoboji the rights of two dozen contemplative perch fishers and two hundred peaceful lakefront homeowners appear to bear the same weight as the whims of a single testosterone-laden lunatic on a jet ski.

Politicians often present us with a simple-minded choice: people or nature (jobs or owls, croplands or wetlands). As nature becomes more distant (deciding to play a compact disc of natural sounds in your house rather than just going outside to listen), it becomes all too easy to dismiss it. And we do not mind destruction if we do not know what we are destroying. It is much easier on a soldier's conscience to fire a bullet at an anonymous enemy than it is to imagine what the bullet does upon impact, not only to the opposing soldier's body, but also to that person's family and friends. What you don't know won't hurt you.

Champions of grass-roots biodiversity must face the realization that in much of the United States, local plants and animals do not stir deep emotions; paddlefish, while having a certain appeal all their own, are not pandas. Nevertheless, I suggest that if you can care about pandas, you can care about paddlefish or amphibians or dragonflies or even clams. Local pride could encompass local natural history, even the most obscure and economically valueless (but, as we will see, not necessarily ecologically unimportant) species. We could even teach local natural history to our grade schoolers.

Community pride in local natural history appears to be a relatively novel idea, but, I submit, one worth test-driving in Okoboji. Okoboji has an abundance of public lands, a recorded account of its native natural history, active protective associations committed to a clean environment, and a biological field station—a source of scientific information. One can argue that if pride in local natural history cannot be embraced by Okobojians, it will not be embraced anywhere. It will have become a concept out of step with these noisy, violent, and destructive times.

America is changing. One important change is that our governments now appear to be less about lives and more about money than has been historically true, at least in my lifetime. This money-based philosophy, which is somehow still uncoupled from fiscal accountability (so let us, this one time, not call it economic-based), is being extended to our natural history. The plant and animal species that can pay their way (or at least get out of the way) will be successful in these modern times, while those that cannot will succumb—money-based selection. (As an exercise, extend this idea to a parallel world, your brain. Imagine a world where the only thoughts worth thinking are those that make you money; then think about your feelings for your children, your friends, your long-time neighbors).

Is this world our future, and, if so, is this what we would like it to be? Recall Mardy Murie's words, in testifying before the U.S. House of Representatives on behalf of the Alaska Lands Bill in 1977: "I hope the United States of America is not so rich that she can afford to let these wildernesses pass by—or so poor she cannot afford them."

Okoboji Wetlands will not be a doom-and-gloom environmental pronouncement. We have choices regarding our environmental policies. All too often arguments over environmental issues become reduced to the rhetoric of extremists. When this happens, metaphors—one of my favorite teaching tools—can help to illustrate the subtleties of discourse. When I was an adolescent toward the end of the Vietnam War, U.S. military history and military strategy interested me. One event in particular that immediately preceded our entering World War II stands out not only as a classic example of wrong thinking but as a metaphor for environmental concerns. At seven o'clock on the morning of December 7, 1941, Privates Joseph Lockard and George Elliot were manning their mobile radar screen at the Opana Station on the Island of Oahu, in Hawaii, guarding the northwestern flank of Pearl Harbor. Radar was new then, and while it had enabled the British to win the Battle of Britain, radar was not completely trusted by the U.S. Armed Forces. When Lockard dutifully reported the unbelievable number of "blips" approaching Pearl from the north, his immediate superiors took what many would consider to be the "conservative" approach. They mistrusted either the private or his machine and did nothing. The rest, as they say, is history: 3,581 American casualties. How much trouble would it have been for the officers in charge to call general quarters, something that must have been done dozens if not hundreds of times previously in mock preparation for such an attack? This would not only have been the wise approach, it would have been the truly conservative approach—the cautious approach. Of course, we survived Pearl Harbor, and the Navy with its aircraft carriers leading the way got its revenge seven months later at the Battle of Midway. But how many sons and daughters were lost unnecessarily at Pearl? Metaphorically, *Okoboji Wetlands* is meant to serve as a Private Lockard. It is our choice whether to listen to and trust the warnings, or not. And while the survival of Okoboji is not on the line, the shape of our future is. Forms of life are at stake here. I reiterate: when faced with environmental deci-

View of the Lakeside Lab grounds in 1945, looking northeast, in a photo taken by W. A. Anderson. From left to right, the labs visible are Macbride, Pammel, and Calvin. Photo courtesy of the Iowa Lakeside Laboratory.

sions, the usual conservative approach—doing nothing—is not necessarily the cautious approach.

The book's organization can be likened to a patchwork quilt. Okoboji's past natural historians—its ghosts—have provided the heritage, the well-worn and comfortable fabric. This information is supplemented with new material written by current and past Lakeside Lab faculty. I then add a border and fit these essays into a pattern. Okoboji's ghosts allow the reader to envision presettlement Okoboji by documenting our historical mammal, bird, and fish fauna. This past world of elk, buffalo, paddlefish, mudpuppies, otters, wolves, cougars, cricket frogs, and prairie chickens was very different from the Okoboji that we currently enjoy. There were few deer and no cardinals in old Okoboji. Who today can imagine being trapped for hours in a willow thicket by thousands of "angry" crows?

Looking toward the future, I then address the wisdom of our current wetland management practices. In particular, I consider the impact of the aquarium ethic, especially its impact on the wetland animals that I know best—our native amphibians. I discuss the role that healthy amphibian populations play in creating healthy wetland environments, then point out the role that healthy wetlands play in maintaining the water quality of our large recreational lakes—water quality so necessary for our summer economic health. Through this linkage I make the subversive claim that native amphibians are our Chamber of Commerce's best friends.

Lakeside Lab itself is a mystery to many native Okobojians, and unknown to most visitors. As an introduction to the Lab, I include essays by two of the very first students. Despite the intervening years, much of what goes on now at Lakeside is similar to what went on in 1910. In fact, the porch where Maude Brown first stood belongs to Main Cabin and is still there. We feel it is appropriate—indeed, it is a metaphor for the way we prefer our natural history—that at Lakeside many aspects of our past remain components of our present.

Throughout this text, the authors' names and the original publication dates of their essays are indicated; full references are presented at the back of the book. The older articles have been reprinted either whole or in part. The more recent, previously unpublished essays are not dated, and unsigned essays are mine. The only changes I have made have been to convert metric measurements to American standard units, to emend spelling and punctuation, and to leave out most scientific names. On occasion, I have inserted a comment within brackets. I have relied on the *World Wildlife Fund Atlas of the Environment* (Lean and others) for data on global environmental phenomena.

Okoboji Wetlands says: this is where we have come from; this is where we are now; this is where we are going. *Okoboji Wetlands* looks back, not for the sake of nostalgia, but to learn lessons about natural history stewardship that may be applied to our future.

Our brains have a second quirk: issues of short-term importance take precedence over long-term perspective. As such, and as we will see, matters of public policy are now often measured against the question of money— how much will this cost me? *Okoboji Wetlands* suggests a new priority, a new measuring stick—will this enrich the world of our grandchildren?

1 AN INTRODUCTION TO WETLANDS

Before we begin, wetlands need to be defined and described. But even before this, wetlands need to be experienced. In the summer, after supper, put on some mosquito repellent, go out to an Okoboji marsh, watch quietly, and listen. Hear the raspy calls of yellow-headed blackbirds defending their territories, get dive-bombed by terns, spread the cattails and look for crayfish chimneys. Watch the damselflies and dragonflies skim the water surface as they patrol for mosquitoes.

Everywhere they occur, wetlands inspire different images—positive and negative—in different people. To the developer, wetlands are wastelands, soggy areas no good for boating or fishing, a breeding ground for mosquitoes. To the biologist, wetlands are wonderlands, waterfowl factories, filters for our larger lakes, buffers against floods and droughts, one of the most productive natural ecosystems on earth. Through education, and reinforced by recent devastating floods, the layperson's attitude is gradually changing from the former perspective to the latter. More than ninety percent (ninety-eight percent in one estimate) of Iowa's wetlands have

been drained for agriculture and urban development. Most remaining wet-
lands are now protected by the state or by private organizations. The
Okoboji region has been soggy for much of the past eleven thousand years.
It remains soggy; despite widespread drainage Okoboji still contains one of
Iowa's highest concentrations of wetlands.

Wetlands defy definition by exact standards, such as acreage or depth, because weather and water conditions vary from year to year. In the prairie
pothole region of the upper Midwest, including the knob and kettle region
characteristic of Okoboji, wetlands are identified by their behavior. Wetlands can be defined as periodically exhibiting either frequent to occasional
drying or deep hypoxia (lowered oxygen levels). Wetlands that never dry or
go hypoxic may rightfully be called lakes. Examples of wetlands in the
Okoboji region include Garlock, Christopherson, Hale, and Jemmerson
Sloughs, and the Gull Point and Spring Run marsh complexes. Welch,
Sunken, Grover, Marble, Hottes, Prairie, Little Swan, and Diamond Lakes,
while termed lakes, may more properly be termed wetlands because they experience hypoxia during periods of drought.

In general, wetlands dry because their basins are small and/or shallow.
Large surface areas and shallow depths promote evaporation, and therefore
rapid drying. Surface areas and water levels are further reduced during
drought conditions. Moreover, small basins with relatively large surface
areas heat up quickly, and warm water evaporates faster than cool water. In
Okoboji, droughts occur in ten-year cycles that center around the ends of
decades. Our last drought years were in 1988–1990, before that in the late
1970s, and so on.

Understanding why wetlands go hypoxic is more complicated than understanding why they dry. Shallow basins do not hold oxygen as well as
deeper basins. One reason is that wetlands hold relatively little water. A
second reason is that wetlands usually support a large number of aquatic
plants (single-celled and filamentous algae, and pondweeds—technically
termed macrophytes) during the summer. Plants photosynthesize—the
process that utilizes carbon dioxide to make simple sugars and produces
oxygen—but only under light conditions, during the day. Day and night,
plants respire—taking up oxygen and producing carbon dioxide—just as
surely as do animals. The combination of nighttime plant and animal respiration depletes the available dissolved oxygen in wetlands. Commercially
important game fish stocked in these wetlands suffocate and die, producing
a summerkill.

Wetlands may also go hypoxic during the winter, producing winterkills, which usually involve fish but may also affect adult leopard frogs and other aquatic animals. Winterkills result from poor light penetration through the surface ice, generally caused by shallow water and a thick snowpack. Sunlight does not reach the water, photosynthesis becomes impossible, and the combined respiration of animals and that of the microorganisms that help decompose dead plants depletes the available dissolved oxygen. Anaerobic (without oxygen) microorganisms then thrive. The end product of this anaerobic metabolism is a buildup of sulfur and methane. Sulfur produces a stench that can be smelled in the spring when the ice breaks up and the water "airs out."

Why is it important to distinguish between lakes and wetlands? Lakes and wetlands are distinct ecosystems based on their characteristics of water permanence and dissolved oxygen levels, as well as temperature. Animals, such as amphibians, that live in wetlands must be able to tolerate the "worst" extremes in these conditions: low water levels, low dissolved oxygen levels, and high temperatures.

I will have succeeded here if I begin to lay the groundwork for an enlightened wetland management strategy, which I define as one based on a thorough knowledge of wetland functions and values. Whether this groundwork will be laid first within the administration of those we have charged with protecting our natural resources or whether it will be laid first within public opinion remains to be seen. I was naively surprised to discover the extent that our public servants have lost perspective on who, ultimately, signs their paychecks. In putting this book together I have tried not to make the same mistake. I will have no complaint if this book is criticized as folk science; although it is based on scientific information, it is not a scientific book. It is written for those people who live in or visit Okoboji and who care about its natural history. People in other regions of the Upper Midwest may also find the information of interest; features of Okoboji's history are shared. I have taught the Vertebrate Ecology and Evolution course at the Iowa Lakeside Laboratory since 1988. (I spent the four preceding summers at Lakeside, and parts of two other summers, dating back to 1977, first as an undergraduate student, then as a graduate student researcher.) As a graduate student in the animal ecology department at Iowa State, I coordinated the first two years of the fisheries management and limnological techniques laboratory course, required for all fisheries biology majors.

Lakeside Lab is the biological field station established for the study of

Aerial view of the Lakeside Lab grounds taken sometime after 1978, after the last of the modern laboratories was built. Note the large number of trees, mostly box elder and ash, now present. Bur oaks, which are resistant to prairie fires, were the predominant trees in presettlement Okoboji. Photo courtesy of the Iowa Lakeside Laboratory.

Iowa's natural history. It is a Board of Regents institution cooperatively administered through the three state universities: Iowa State University, University of Iowa, and University of Northern Iowa.

Lakeside was founded early in the postsettlement history of Okoboji—in 1909—and has provided an extensive scientific record of the natural history of our area. The members of the Okoboji Protective Association—founded at about the same time—have also generated a great deal of natural history information and have expressed many concerns about the environmental health of our region.

The following pages gather and examine much of this material, especially information recorded before 1930, intertwined with my own more recent observations and those of my Lakeside Lab friends. There is much for the reader to discover. For example, you may wish to know what you probably suspect, that the natural history of our region is not the same now as it once was. You may want to learn the identity of the animals that were once here and are now gone, and, conversely, which animals are exotics. You may find it amusing that locals were calling the Lakeside Lab the "bug house" as far back as 1910, a mere year after it was founded. You may not be surprised to learn that we have some environmental problems, most of which can and should be corrected. If we have the will to correct them, the overall message is one of guarded optimism.

For the purposes of this book, I define Okoboji geographically as all of Dickinson County, Iowa, from Milford north, including Milford itself and Milford Creek. Thus, Okoboji includes East and West Okoboji Lakes, Big and Little Spirit Lakes, Silver, Diamond, Welch, Center, Grover, Little Swan, and Prairie Lakes, the east and west forks of the Little Sioux River and the Little Sioux itself after the confluence, Upper and Lower Gar Lakes, Minnewashta Lake, Jemmerson, Garlock, Hale, Henderson, Yager, and Christopherson sloughs, the Kettleson Hogsback and Spring Run wetland complexes, Silver Lake Fen, the Excelsior Fen Complex, Dugout and Pillsbury Creeks, Cayler Prairie, the Freda Hafner Kettlehole, Gull Point State Park, the Lakeside Laboratory grounds, and the many prairie potholes and upland natural areas scattered throughout this region. (For the locations of these public lands, I refer readers to Bohumil Shimek's 1920 map of Dickinson County, included here, or more recent maps: Recher's *Iowa Great Lakes Recreational Map*; Harr, Roosa, Prior, and Lohmann's pamphlet, *Glacial Landmarks Trail: Iowa's Heritage of Ice*; or the *Iowa Sportsman's Atlas*.)

Okoboji's landscape is what geologists call a terminal moraine; it represents a specific geological occurrence—the western front of the Altamont advance of the Des Moines Lobe of the Wisconsin Glaciation, which ended just north of what is now Milford over one hundred centuries ago. Milford sits on the weathered outwash of this glacier. Today, gravel excavation is one of Milford's leading industries.

In the post–Civil War era, Frederick Douglass urged young people fighting for their rights to "agitate, agitate, agitate." My own motto for conservation, paraphrasing Douglass, is to "educate, agitate, educate," thus providing an admonishment, the second meaning to this book's subtitle (the first meaning, of course, is instructional). This is an arrogant but necessary position, as you will see, in the face of bureaucratic rigidity—in the face of the "aquarium ethic" underlying the management policy currently being applied to our deeper wetlands. (A recipe from the Aquarium Ethic Cookbook: start with fish, add processed chow, mix in some oxygen, apply biocides, and, voilà, one has a supposedly viable ecosystem. After all, what works for an aquarium must also work for a wetland.)

In taking my stance, I advocate a conservative position. By this I do not mean the moral conservatism so politically fashionable in our country today. Rather, I support the conservatism of Teddy Roosevelt and Iowans

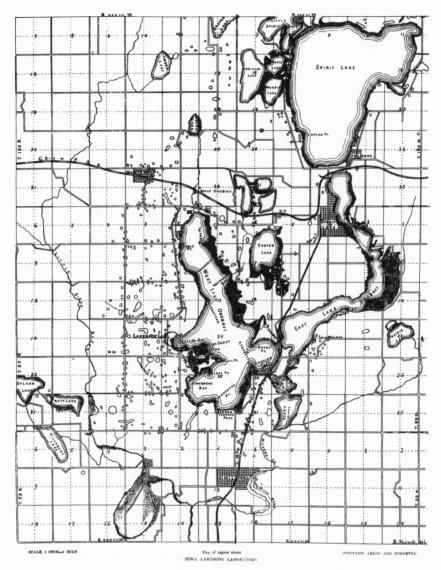

Map of region about
IOWA LAKESIDE LABORATORY

B. Shimek, del.

Bohumil Shimek's 1920 map of Dickinson County. Note the high concentrations of wetlands indicated in the region to the west of Lakeside Lab and along the Burlington, Cedar Rapids, and Northern Railroad tracks from Milford north to Okoboji. These wetland densities reflect their abundance throughout the region. Shimek's access to wetlands—and therefore his depiction of them—was limited by the transportation options of his day. Note also the presence of Sylvan, Pratt, and Pillsbury Lakes in the southwestern corner of the county. This huge wetland complex has since been drained for agricultural use, although evidence of these basins, their shorelines, and their animals remains today.

Aldo Leopold, Paul Errington, and Ding Darling—conservatism in the sense of conservation; conservatism independent of short-term economic motives. By this definition, conservatives could never advocate destroying a wetland, by whatever means or for whatever purposes. Traditional American values should include traditional American animals and their ecosystems. My own moral stance is epitomized by Aldo Leopold's famous land ethic, presented in *A Sand County Almanac*: "A thing is right when it tends to preserve the integrity, stability, and beauty of the biotic community. It is wrong when it tends otherwise." This is a conservative position. This is conservation.

Currently, there are about two dozen experts on the natural history of Okoboji. One possible format for a book such as this would be to muster all of today's talent and present an in-depth snapshot of our current environment. As important as this would be, I have not done this. Instead, I have chosen to examine the history of the changes in Okoboji's natural history by turning to Okoboji's ghosts—the spirits (in both the metaphysical and vivacious senses) of our past natural historians—to help me record the history of Okoboji's wetlands and their animals.

Amphibians will be emphasized. Why amphibians? One reason is that amphibians are among the most important groups of game animals in the history of Okoboji. Although mostly forgotten now, the economy of turn-of-the-century Spirit Lake hinged on the collection of tens of millions of adult northern leopard frogs. Spirit Lake frogs provided the frog's legs served in restaurants throughout the Midwest and as far east as Philadelphia. Second, I have discovered some things about Okoboji's amphibians. My own scientific research, conducted in Okoboji over the past sixteen years, has focused on wetland ecology and the role amphibians play in wetland ecosystems. Did you know that Okoboji has a cannibalistic form of tiger salamander that has never been found naturally anywhere else in the world? If the University of Okoboji is looking for a new mascot, cannibal morph tiger salamanders would get my vote. A third reason for highlighting amphibians is that these vertebrates are known to be sensitive indicators of environmental quality. Amphibians do not have protective body coverings such as hair or feathers, and they spend part of their lives in the water and part of their lives on land. Amphibians need good quality aquatic *and* good quality terrestrial habitats if they are to survive. In Okoboji, amphibians are dependent on natural wetlands to a much greater degree than any other vertebrate group, including waterfowl. Amphibian health is a measure—a proxy—of wetland health.

Okoboji's ghosts allow me to present a chronology of not only where we are today but how we got here. This solves one problem: it gives us a past. Gretel Ehrlich, in *The Solace of Open Spaces*, has said: "We live in a culture that has lost its memory." Some may think that there is nothing wrong with this. "What the kids don't know they won't miss," says a Kansan on the demise of the tallgrass prairie in William Least Heat-Moon's *PrairyErth*. People who embrace this position run into a difficulty: the old timers could write and our kids can read.

But there exists an even more damaging opinion, from the same paragraph in *PrairyErth*: "What the goddamn hell are they [the Kansas Fish and Game Commission] doing putting antelope back in here for? Everything we worked a hundred years to get rid of, they're bringing back." This is frightening and wrong—biologically wrong—for the following reasons. First, we must advocate policies that promote the conservation and diversity of our native plants and animals, and oppose policies of exotic species introductions. Why? If for no other reason than we know that native species in natural communities form interactions that work. These species coexist and in many cases have coevolved. Native species form self-sustaining communities that are naturally productive. The rich, black Iowa soil that fuels our economy did not form by accident, and it is not continuing to be produced at sustainable rates. Our current soil fertility is the result of the natural biology of our native prairie plants and the diverse communities that they once formed. Healthy communities that for thousands of years laid down rich layers of organic material without the addition of fertilizers and pesticides. (A lesson: go into a freshly plowed Iowa field and turn over a big dirt clod. Notice something missing? What happened to the earthworms? Did you know that earthworms were responsible for making the topsoil in the first place? One is reminded of the old Will Rogers quip: "They're makin' people every day, but they ain't makin' any more dirt.")

Exotic species introductions rarely fit so nicely into existing communities. Often, introduced species either die out (in my view the best-case scenario) or completely take over. There is a premise in ecology that diversity equates with community health and stability—that is, the more species that compose a community, the more interactions can and will occur. But exotic species first displace and then replace native species. (When exotic bullfrogs colonize an Okoboji wetland, this one species displaces four species of native amphibians.) Community stability is threatened as these effects reverberate. Imagine your car's engine as a biological community. As it is—in its native, factory-installed state—it runs well. Now, replace one species—say, the distributor cap—with another species—a different type of distributor cap. Does the distributor work as well? Probably not. And notice that not only is distributor function affected but also spark plug wire function; and through it, piston function; and through it, drive shaft function; and through it, engine function. Not only have we lost the original distributor cap species, we have lost the engine community. The car is broken.

Second, we should take the concept of native species and carry it one step further: species that are found elsewhere in Iowa but not historically in Okoboji are exotics; they are not our native species. Political boundaries are not biological boundaries. Biologists who promote statewide distributions of regional species lose their credibility as ecologists and instead become a type of politician. But think about it, politicians have power. Which trait is more critical to an individual in today's society, power or credibility? Being a scientist, I might choose one, whereas some might choose the other. This may be the root of Okoboji's wetland problems. Ding Darling consistently opposed Iowa's policy of staffing the top positions of its Department of Natural Resources with political appointees to no avail (see Lendt 1989). Of course, Iowa is not alone in politicizing its natural resources positions; there are federal precedents. Recall Ronald Reagan's 1981 appointment of James Watt as U.S. Secretary of the Interior. Before that, in 1952, Dwight Eisenhower appointed automobile salesman Douglas McKay to the same position. As Rachel Carson (the author of *Silent Spring*, the influential book warning of the dangers of pesticides) wrote in *Reader's Digest* in response to McKay's appointment: "The real wealth of the Nation lies in the resources of the earth—soil, water, forests, minerals, and wildlife. To utilize them for present needs while insuring their preservation for future generations requires a delicately balanced and continuing program, based on the most extensive research. Their administration is not property, and cannot be a matter of politics" (Brooks 1989).

Third, we must realize that economic (in the usual sense) gains that run counter to long-term environmental health cannot be sustained. People today have been taught to view environmental concerns as a drag on our economy. Instead, we need to view our economy as being nested inside our environment. It is a fact that our economy comes from our environment—without our environment we would have no economy. We do not farm in a vacuum. As our environment deteriorates (where did those earthworms go?), eventually so must our economy. Maybe this will not happen in the short term—there is a lag time—but history teaches us that it will happen in the long term. Seeing Iowa's topsoil erode, and knowing that the organisms that create topsoil have declined, what can we conclude about the long-term future of Iowa's soil fertility, and therefore the future of Iowa's economy? We only need to look to other places to know. Mesopotamia, the cradle of civilization, the so-called fertile triangle, has been transformed by agricultural overuse into a desert. Try to sustain a first-world economy in a desert. "Only to those who have seen the Mesopotamian desert will the evocation of the ancient world seem well-nigh incredible, so complete is the contrast between past and present . . . it is yet more difficult to realize . . . that the blank waste ever blossomed, bore fruit for the sustenance of a busy world. Why, if Ur was an empire's capital, if Sumer was once a vast granary, has the population dwindled to nothing, the very soil lost its virtue?" (Ponting 1936).

There is a state-supported policy of natural history (mis-)management now being applied to Okoboji's deep wetlands. I have taken to calling this philosophy the "aquarium ethic." This term is based on Aldo Leopold's land ethic. Leopold advocates developing a set of principles—of values— for treating and caring for our land. While Leopold's land ethic teaches us to work with, and to understand, the earth as we go about managing it for human uses, Okoboji's aquarium ethic presents a contrasting approach to wetland management. Under the aquarium ethic, any management plan for deep wetlands that might be derived from an understanding of what wetlands are and how they function is secondary to their perceived role as vessels to raise game fish. My descriptor is therefore not complimentary, nor should it be. To be sure, the aquarium ethic is not an overt policy of active wetland extirpation. It is simply a game management policy that emphasizes fishing license revenues over native natural history. Of course, the rationale underlying these practices does not matter to wild plants and animals, these organisms only experience the result—extirpation.

The current aquarium ethic involves the use of two types of poison, one of which persists in our waters for up to a year. It also involves connecting fringing wetlands to our larger lakes in the unproven and, in my experience, mistaken assumption that this will increase gamefish spawning habitat. And the aquarium ethic involves the introduction and spreading of non-native bullfrogs into our largest wetlands. (Bullfrogs are currently not being stocked, but nothing has been done to eliminate them either, and existing populations are expanding.) This introduction has been done without regard to what impact these exotics will have on our native amphibians or, in fact, on our wetlands.

The aquarium ethic has affected all of Okoboji's large wetlands. In advocating native natural history over the course of this book, I will challenge this management program. But understand that by challenging game fish management practices, I am not challenging Okoboji's game fishing industry. Isaac Okoboji need not feel threatened. The problem of providing good game fishing has several solutions. Is Okoboji's the best? As this book proceeds, I will leave it up to you to decide.

2 ORIGINS

To know a thing fully, you must understand something of its history. Okoboji's written history begins with the appearance of European settlers a century and a half ago. Its current landforms appear much earlier, with the receding of the Wisconsin Glacier 11,000 years ago. Our modern understanding of Okoboji's natural history and its ecological relationships starts with the establishment of the Lakeside Laboratory in 1909. The following three essays detail these origins, allow us to view an Okoboji altogether different from today, and in the process begin our story.

ABOUT THE AUTHORS

Abbie Gardner Sharp is famous in Okoboji, the teenaged girl kidnapped for 83 days in 1857 by the Lakota Sioux warrior Inkpadutah after he and his band killed the rest of her family and all but one of the other settlers at Spirit Lake. After the state of Minnesota paid a ransom, young Abbie was released and returned to Okoboji. Inkpadutah's strategy of scaring off settlers backfired. As a result of the publicity, Okoboji was quickly repopulated by about 900 new settlers, and elections were held soon after.

Abram Owen Thomas was a professor of geology at the University of Iowa. Born in Wales, Great Britian, in 1876, he came to Iowa six years later. Thomas received his master's degree in 1909 from the University of Iowa and was such an effective teacher that he was retained. Thomas received his doctoral degree in 1923 after a year of intensive study at the University of Chicago. He became a member of the Iowa Academy of Science in 1904 and was its treasurer from 1915 until he died in 1931. About his teaching, one student said: "When Professor Thomas appeared before his classes and spoke with his dry Welsh humor and homely farmland phrases, the drab academic cloak of his subject fell away and his hearers caught the fire and inspiration of one who bowed before the majestic grandeur of the universe."

Thomas H. Macbride (1848–1934) is the founder of the Iowa Lakeside Laboratory. Macbride was a distinguished member of the University of Iowa faculty and its president from 1914 until 1916, the year he retired. He is perhaps best known in Iowa today for a structure he probably would not have approved of— the reservoir that bears his name (which is now misspelled).

SETTLEMENT
Abbie Gardner Sharp, 1908

Although half a century or more has passed away since then, memory still recalls that sultry day in July 1856, when the white covered wagons heavily loaded with household goods and provisions, and drawn by ox teams, brought the first white families to the shore of this far-famed beautiful lake region. Seldom ere this had the numerous beauties of these lakes greeted the eye of a white man. Their waters had slept for centuries unknown to the turmoils of civilization, disturbed only by their finny inhabitants, wild fowls or the rippling oars of the Indian canoe.

Great schools of perch, bass, pike, pickerel, and buffalo fish had long gamboled below their transparent waters without fear of the white man's hook. Every variety of wild fowl, such as geese, ducks, pelicans, prairie chickens in flocks which no man could number, flourished here in all their native loveliness; likewise the swan proudly curved her neck as she floated her snowy bosom over the deep waters, exulting in a realm where she reigned supreme and sole monarch. Herds of elk and deer, in all their native freedom, roamed over the prairies and fed on the nutritious grasses and sought shelter in the shady groves. Wolves [likely coyotes] were also found here in abundance and the nights were frequently made hideous by their howling around the log cabin doors. My father, Rowland Gardner, was an ardent lover of nature. Especially fond was he of a lake region. Amid such charming scenes as these and with bright prospects for the future before him he felt that he had found the "promised land" and that here on the shores of Okoboji he might settle down and spend the evening of his declining years in a peaceful home surrounded by his family.

THE GLACIAL STORY OF THE LAKE OKOBOJI REGION
A. O. Thomas, 1913

The Okoboji region with its variety of topography, its chain of lakes, its location near the highest land in our state, its interesting drainage problems, and the charming story of the origin of it all, affords endless pleasure and profit to a student of the lake district. Few, indeed, are the visitors who do not admire the setting and who do not marvel at the natural forces that have been at work developing this fascinating topography. Even the casual tourist is impressed with the unusual arrangement of the knobs and basins as well as with that of the lakes themselves.

Tens of thousands of years ago a field of snow and ice occupied a large area west of Hudson Bay. A similar field lay in Labrador. The snows of successive winters added to their masses, while the summer sun did little more than partly to melt each season's accumulation. The melting of the surface snow and the sinking and refreezing of the water below the surface gradually changed the snow to solid ice. Little by little as the ice-cap grew in extent and thickness, it acquired slow motion outward from its center much as does tar when poured upon one spot on a flat floor. Both ice fields spread in this very slow but irresistable manner, at last uniting into one field in the Great Lakes region. This field deploying southward covered at the time of its maximum extent that part of North America stretching from the Canadian Rockies to Labrador and from Hudson Bay southward nearly to the latitude of Cincinnati and St. Louis. This continental ice cover was thick—how thick we shall not say, but judging from a similar moving field which covers Greenland today, it must have been hundreds, perhaps thousands, of feet in thickness.

The momentum of such a mass is beyond comprehension, and as it flowed along all irregularities of the landscape were overridden. The lower part of the ice soon became filled not only with boulders large and small but with pebbles, gravel, sand, clay, and so on. The differential movements of the ice caused this englacial detritus to be partially ground up in transit while the rock-shod ice planed and scoured the country over which it passed. Sooner or later the coarse and fine were dropped, some to fill hollows and valleys, some on flat areas when the load was too great, while some was carried the whole journey, being finally deposited along the flanks or front of the glacier as it wasted away. This transported material, known as drift or glacial drift, is often scores of feet in thickness and composes the chief part of the soils of our state. The boulders, often grooved and polished, form conspicuous and attractive features of the lake region; almost every variety of rock is represented among them and the parent ledges from which they have been brought are far to the north. A stroll along the lake shore will repay the visitor; here he may see "walls" of these boulders—granites, traps, gneisses, schists, quartzites, and so on in endless profusion; the waves lap them, the winter's floating ice buffets them, yet this is all as a lullaby compared to the stress and strain of the mighty glacier which carried them from their northern home.

When at last the icy mantle withdrew under the influence of a more genial climate, and when the land surface was laid bare, the once established stream courses, the hills, and the hollows had been effaced. Only a rolling plain with low broad sags or gentle swells broken here and there by rounded knobs and shallow water-filled basins remained. Plants and animals gradually reestablished themselves in the wake of the melting ice. Running water once more had to carve a system of drainage to replace the one obliterated by the glacier's occupancy. A genial climate lasting thousands of years passed by when again the northern lands became the abode of an ice field and again a large part of North America was covered with a glacier; this in turn retreated and was likewise followed by an interglacial interval. Five times in all did this recur; no two of the glaciers covered exactly the same territory; the first and second crossed Iowa, the third barely entered the state from the northeast across the Mississippi river . . . and [the] last developed a broad tongue-shaped lobe which extended into the state as far as the site of the present city of Des Moines; its maximum width along the northern state boundary was from Osceola county east as far as Worth county.

In imagination let us for a moment approach our lake region from the southwest at the time of the maximum extent of the last glacial lobe—known to geologists as the "Wisconsin lobe." The western flank of this great lobe passed diagonally across Dickinson county and as we draw near a great field of ice with an irregular, jagged margin like a bold coastline breaks upon our view—if in winter, the view is a most majestic one, for the glacier is now snow-covered and glistening white [we now know that glaciers supported soil and vegetation]; in the background like a low mountain range the glacier extends with a northwest-southeast sweep along our horizon; if in summer, great floods of water spread out towards us, choking up every depression with such glacial detritus as water can carry but flowing on in a sheet to the west and south and seeking valleys less filled because of their remoter distance. The surplus drift along the margin is being piled up into irregular heaps and ridges, technically called moraine. The ice, gaining in plasticity by the melting, is flowing slowly outward and in so doing it plows up the moraine into still greater ridges and knobs.

For a long time the annual melting and the annual advance were practically equal; this fact kept the glacier from gaining any new ground and compelled it to work and rework again and again its belt of moraine. With a further gain in annual temperature the retreat began and the front of the ice melted away farther and farther to the north and east until at last it disappeared from the state and eventually from the continent.

It is evident that glaciers on the whole smooth out the topography and decrease the relief of the country over which they pass; the level stretches so common in counties to the south and east of the lake region are good examples of such a topography. On the other hand, the margin of a glaciated area has normally greater relief, that is, is rougher, than it was before the glacial visit. In such a rough morainic area lie our lakes. In fact the belt of moraine surrounding the Wisconsin lobe contains all the large lakes of our state, Spirit Lake, the Okobojis, Storm Lake, Wall Lake, and even Clear Lake a full hundred miles away in the moraine of the eastern side of the lobe; scores of smaller but beautiful bodies of water similarly situated might be mentioned, such as Elbow Lake, Rush Lake, Silver Lake, Lost Island Lake, and so on, while countless unnamed pools lie nestled among the knobs and ridges. The knobs, some of them with local fame as "high points," are simply the counterparts of the lake depressions, while the tough glacial clay out of which the basins were scooped are too impervious to permit the loss of their waters by underground drainage. Moreover, our lakes are located not only in one of the roughest parts of the moraine but are also near the crest of the divide between the Mississippi and Missouri rivers. All these conditions make for the perpetuity of the lakes and since there are no permanent streams emptying into them their natural preservation is assured for ages to come. Man should and will do his part in conserving their virgin beauty in its natural setting. Our occupation of the region is but a fraction of the countless centuries since the lakes were so wonderfully made. It is our duty to transmit their clear blue waters unsullied to generations yet unborn. [For more information about Okoboji's geology, and its relation to Iowa's geology, see Jean Prior's *Landforms of Iowa*.]

THE OKOBOJI LAKESIDE LABORATORY
Thomas H. Macbride, 1909

The establishment of the Okoboji Lakeside Laboratory, founded by the alumni of the State University of Iowa, promises to affect so deeply the future scientific work of our state that some account of its beginning and especially its *raison d'être* may rightly claim the attention of the [Iowa] Academy [of Science]. The laboratory has been located on the west shore of Lake Okoboji in Dickinson county for the reasons following:

In the first place the topography of Dickinson county is peculiar, unique. Situated on the western border of the Iowa Wisconsin drift, the region illustrates, as possibly no other equal area in the state, the special characteristics, not only of glacial moraines in general, but in particular the very expression of the Wisconsin moraine. In fact, I think that it must be admitted that the Okoboji lakes and their encompassing hills do indeed form the finest bit of morainic topography to be found on our western prairie.

This fact, of course, makes the locality an especially interesting field for illustrating to the student all the fascinating features of the latest page in the

Macbride Lab under construction by the Civilian Conservation Corps, 1936.
Among the clues that identify this building is the unique pattern of shades
and shapes of its stones. In the background to the right, note that the natural
absence of trees permits views of Little Miller's Bay, the sandspit, and across
West Okoboji to Arnold's Park. Photo courtesy of the Iowa Lakeside Laboratory.

geologic history of our state. Indeed, the very fact that the locality is mar-
ginal makes it especially interesting, and studies of contact, of movement
and retreat, as well as of direction and relation to pre-existing topography—
all these things are especially accessible and patent within half a day's drive
along any highway south or west.

Secondly, the region having Okoboji for its center is, by reason of the peculiar topography just mentioned, the field of a special floral display
difficult to illustrate anywhere else within such narrow limits. We have a
forest flora and a prairie flora; and neither in this part of the world has ever
been adequately studied. It is believed that the fungal flora of the region, for
instance, is especially rich and interesting. We have all kinds of habitat conditions, from aquatic to xerophytic. We have deep water, shallow water, permanent [water]; marshes, springs; and xerophytic slopes and hill-tops, some
so dry as to offer home to the vegetation of the higher western semi-arid
plains. The plankton of the lakes is filled with desmids and diatoms and all
manner of algal flora, during July and August rich beyond comparison in all
that makes up the tide of life for these simple but fascinating forms.

Photographic postcard of Macbride and Shimek labs and the front gate of Lakeside Lab. This postcard was sent from Lakeside to Martin Grant, long-time professor at Lakeside, by J. R. Murray. Photo courtesy of the Iowa Lakeside Laboratory.

Neither have the xerophytes [upland plants] been studied nor the flora which joins these, perchance, in ecologic bonds with their aquatic congeners, for the waters are filled with flowering plants, richly indeed as with floating cryptogams, and the factors of ecology and distribution are all here, in large part so far, unexplored and certain to interest for centuries generation after generation of Iowa students.

For similar reasons, the fauna of the lake district will reward our constant study. The varied flora, just described, insures a varied fauna. The waters teem with animal life. Probably the protozoa of the whole valley will be found hiding on the vegetation of these quiet lakes and pools. Of course, the avian and vertebrate aquatic fauna are rich, and even the terrestrial vertebrates are likely to prove more than commonly worthy of investigation. While this is [in] writing the papers tell of a mountain lion shot in one of the near-by marshy lakes! It is not believed that carnivores of size are likely to abound, not to such extent at least as to warrant a future visit from our nimrodic ex-president [Teddy Roosevelt], but it is believed that natural science, in all its branches, entomology, ostracology, ornithology, will be greatly enriched by using such opportunity for research as Okoboji may afford.

The people of Lakeside Lab in 1909, the first year of its existence. Thomas Macbride is in the center. Note the formal dress. Macbride's crew is posed in front of Main Cottage. Photo courtesy of the Iowa Lakeside Laboratory.

Again, Okoboji as the world knows is already a place of resort; thousands of people find summer habitation on its shores. So that we find here unequaled opportunity for bringing scientific work to the attention of people of every class and kind, and confessedly natural history work in all our schools, colleges and universities is too formal, too artificial, too much based upon material specially prepared, laid up in herbaria, or conserved in cases and bottles; the Lakeside Laboratory offers an opportunity to correct this, at least in some small measure. Ever since the immortal Agassiz stood bareheaded with that famous company on the rocks of Pennikese, the naturalists of the world, at least, have realized that the proper and reverent place for the study of natural objects is in their natural surroundings. Dry dead fungi are dusty labelled things, as meaningless as the stuffed skin of mammal or bird, or a fossil in a box; better than no exhibit at all, to be sure, but poor indeed as compared with the natural world, where the fungus starts in the forest shade, the wings of bird or insect fan the sunny air, or the fossil speaks its significance from the stony pages of the riven quarry stone. The Lakeside Laboratory shall afford to all interested, for once at least a chance to see the real world, nature alive, accomplishing her miracles in their own silent splendor, often needing not, for the student's appreciation, the voice of interpreter or teacher.

View of Lakeside Lab looking west from the Miller's Bay sandspit in 1909. The main cottage is to the left; the laboratory building, which no longer exists, is to the right. Photo courtesy of the Iowa Lakeside Laboratory.

A few words now may describe the provisions making for natural history work. The university alumni have purchased property for a plant. About five acres of ground with a cottage for administration purposes, a boat-house, pier, and so forth, are already the equipment. It is expected within the next few days to erect a building for laboratory uses. This building will offer office, library and laboratory for each professor in charge of a line of work. A large hall to seat 125 people comfortably will be accessible for general lectures, evening entertainments, and so forth, and from the university such apparatus will be supplied as to enable ordinary classes to work successfully in botany, geology, and zoology. Boats and dredges also will be at hand, while public conveyances enable students to reach conveniently more distant points of interest. It is proposed to offer tents and cottages to all comers up to the limits of laboratory accommodations; at present a class of not more than thirty is in contemplation; and simple meals will be obtainable at reasonable rates.

The classes sought to be accommodated are: first, all students of nature competent to enjoy the laboratory method of instruction; the laboratory shall be open to anybody capable of using its privileges; second, teachers of biologic subjects in academies and high schools everywhere, who may desire to combine recreation with work and who may find in the service of the laboratory occasion to acquaint themselves with Iowa conditions and thus better equip themselves for serving the children of the schools; third, graduate students who may desire to perfect themselves in some line of research preparatory or introductory to an advanced degree. Such students are pre-

The inside of the laboratory building of Lakeside Lab in 1909. Photo courtesy of the Iowa Lakeside Laboratory.

sumably competent to conduct work for themselves, needing simply a place at the table and such suggestions as occasion may develop.

In general, however, the laboratories for use are to be open to the world, students enrolling in the order in which application arrives. It is hoped that the open door may be thronged and that the enterprise may not only serve those already engaged in scientific work, but may reach and influence thousands and make real all natural science to the upbuilding and quickening of every school, college or academy within the borders of our state.

3 · ECOSYSTEMS

Water permits life. While the focus of this book is on wetlands and their animals, wetlands are not isolated ecosystems. The water in our wetlands comes from rainwater and snowmelt, either directly or through runoff, and to a lesser extent from groundwater flow. The water from our wetlands that does not evaporate flows as groundwater (or these days through culverts) into our lakes and eventually into the Little Sioux River. To fully appreciate wetlands, we must consider first their water source—rainwater and runoff—and last their water sink—our big lakes. The following essays address rainwater runoff on Cayler Prairie, characteristics of wetland vegetation, lake vegetation, a historic summerkill in Upper Gar Lake, West Lake Okoboji's annual temperature cycle, and the human impact on West Okoboji, which this lake has recorded itself in its bottom sediments.

ABOUT THE AUTHORS

Richard Baker and Kenneth Lang are current members of the Lakeside Lab faculty. Baker teaches the geology course; Lang, the aquatic ecology course. Both professors continue the tradition of teaching and research excellence that has characterized Lakeside Lab from its earliest days. As a personal note, I count these two men among my very closest friends.

Robert B. Wylie (1870–1959) received his Ph.D. from the University of Chicago in 1904; before that, he taught at Morningside College. In 1906 he joined the University of Iowa. Wylie was the director of Lakeside Lab from 1919 to 1923, chairman of the botany department at Iowa from 1919 until his retirement in 1940, and president of the Iowa Academy of Science in 1922–1923.

M. J. Lonergan was an assistant sanitary engineer for the state of Iowa and does not appear to have lived in Okoboji. According to the 1930 directory published in the OPA Bulletin, *he did not own a cabin or have another business in the region.*

RUNOFF ON CAYLER PRAIRIE
Richard Baker

The summer of 1993 will be remembered for a long time. It was, of course, the wettest summer on record, the summer in which all regions of Iowa suffered the effects of the flood at one time or another. In Okoboji, we had four inches of rain within three hours, and several other downpours of more than two inches—all on a saturated landscape. Runoff was rampant, causing extensive property damage, soil erosion, gullying, and crop loss.

It was instructive to go out to Cayler Prairie, the state preserve located two miles west of Lakeside Lab, the day after each major rainfall. Although the water level in the potholes on the prairie rose slightly, there was no runoff and virtually no soil erosion or gullying. These incredible rainfalls were absorbed by the prairie soil, the prairie plants, and the wetlands. A striking comparison is available with the adjacent land. In these row-cropped lands, our recent rains caused extensive gullying and coincident large-scale soil loss. In fact, the long-term effects of cultivation on soil loss can be measured: at the preserve fence lines along hilltops the soil of Cayler is more than a foot higher than that of the surrounding croplands.

Clearly we can never go back to a time when prairies covered the state.

But it is useful to realize that the flooding problems of 1993 are a modern phenomenon, never realized before human settlement. It is also instructive to realize that, in terms of Dickinson County soil conservation, it could have been much worse. In 1993, slightly over twenty-five percent of Okoboji cropland—the maximum percentage allowable—was in set aside acreage under the Conservation Reserve Program. In 1993 this conservation—this conservativism—paid off.

WETLAND VEGETATION

PLANT ZONATION

Prairie pothole wetlands look distinctive because of their circular plant zonation, which is caused by water depth. Three plant zones—emergent, submergent, and floating—not only characterize wetlands but define the regions that in large part determine where wetland animals live.

EMERGENT VEGETATION

To germinate, emergent plants such as cattails, rushes, and sedges need the mudflats produced by dry conditions. Subsequent flooding does not affect them. This is the zone of marsh birds—red-winged and yellow-headed blackbirds and marsh wrens. It is also the zone of dragonflies and damselflies. Upland animals coming for a drink enter the emergent zone. Amphibians exiting the marsh following metamorphosis pass through this zone.

SUBMERGENT VEGETATION

Submerged plants thrive in wetlands, where shallow depths permit sunlight to penetrate to the bottom throughout the basin. Plants such as coontail (*Ceratophyllum demersum*), the native water milfoil (*Myriophylum exalbescens*), bladderwort (*Utricularia vulgaris*), ribbon weed (*Vallisneria americana*), and the true pondweeds (many species in the genus *Potamogeton*) grow rapidly. Later in the summer, as these submergents grow and as water levels drop, wetlands may become completely filled with vascular plants. At this time the stems and leaves of submergent plants may come to lie on the surface of quiet waters. This may be unsightly or even a nuisance to boaters and to fishers, but it is natural. This is the zone where many of our waterfowl feed.

FLOATING VEGETATION

On certain marshes, the numbers of tiny floating plants, called duckweed (*Lemna minor*), may be so thick that they limit light penetration and retard the growth of submergent plants. Usually however, duckweed rafts are at the mercy of the wind, piling up on the leeward shore. Sometimes duckweed is called pond scum. But duckweed is a tiny vascular plant with roots, leaves, and flowers. It has none of the slimy texture of algae, the stuff of real pond "scum." Duckweed is not misnamed; this is the zone where our waterfowl nest, rest, and seek protection.

LOW OXYGEN LEVELS

The irony of wetlands is that while we are used to thinking of plants as oxygen producers through photosynthesis, nighttime plant respiration draws dissolved oxygen back out of the water to produce conditions of low oxygen (hypoxia). This process of course occurs in lakes also but is so much more apparent—the effect so much more dramatic—in wetlands because, compared with lakes, wetlands have proportionately higher levels of plant biomass. Wetland animals, unlike lake animals, need to be able to cope with low dissolved oxygen levels. For details of the vegetation characteristics and other aspects of pothole wetlands, including Okoboji's, I refer readers to Arnold van der Valk's *Northern Prairie Wetlands*.

THE AQUATIC GARDENS OF OKOBOJI
Robert B. Wylie, 1912

No feature of the Okoboji region should be of greater interest to the summer visitor than the water plants of these lakes. While in the main their waters are clear and deep, and relatively free from plant life, there are shallow bays and shelving shores bearing a rich aquatic flora. To many of the cottagers these are only "water weeds," and their presence, if noted at all, is passed over lightly. Others are made conscious of them only when spoon hooks become entangled in their branches, or the propeller of the launch is muffled by their stringly stems. On the other hand, some are coming to appreciate the important part these plants play in the balance of life in the lakes and as such realize that bass and pickerel would be banished from their haunts with the destruction of these "weeds."

The submersed fresh-water vegetation falls chiefly into two natural divisions, differing as greatly in structure as in history. The simpler plants are called algae; this group includes those forms that have always lived closely associated with water. The other and more conspicuous group is made up of flowering plants which have moved into the water more recently from the land. Representatives of these two widely different classes dwell intermixed, and often in the closest association, in our lakes and ponds.

Most of the algae are small, many of them of microscopic size, though they may develop in immense numbers when conditions are favorable. The yellowish tinge of the waters of East Okoboji Lake in midsummer is due primarily to the myriads of tiny, one-celled algae—diatoms—present in the water. Others of such free-floating forms are larger, and may accumulate in considerable masses at the windward side of open water; some of these are conspicuous in Center Lake in July and August.

In sharpest contrast with these simple algae are the submersed seed-plants which form the conspicuous element of the aquatic flora. Immense beds of these may be seen in the corner of Miller's Bay near the Lakeside Laboratory, in the Gar Lakes, at the upper end of East Okoboji Lake, and in the numerous shallow lakes and ponds of the region. With the exception of the tiny floating duckweed, these are large plants, and possess leaves, stems, roots, and flowers, though the blossoms are often small and inconspicuous.

REPORT ON NUISANCE CONDITION, UPPER GAR LAKE
M. J. Lonergan, 1930

Upon arrival at Arnold's Park on the morning of October 10th a very decided and very bad "pig pen odor" was evident at the post-office, which is one-half mile from Gar Lake. It was a very easy matter to trace this odor down to its source.

Mr. J. D. Hardiman, local Game Warden, stated that the odor first started at Gar Lake on October 2nd. From this time on the odors seemed to get worse and at the time of our visit on October 10th he stated it was not quite as bad as it had been a few days previous. A peculiar thing in connection with the odors was that the fish began to die at about the same time the odors were given off. Game Warden Hardiman estimated that approximately 500 pike of about three pound average were killed and that thousands of fingerlings of all kinds had also been destroyed. He employed three men who spent three entire days burying these dead fish along the shore of Gar Lake.

A very peculiar thing in connection with this fish killing affair was that the fish died in no other lake except Upper Gar Lake. There was some slight trouble at the narrows on about September 29th. The narrows is a short and narrow neck of water in East Okoboji Lake and is located just south of the town of Spirit Lake. This is one of the shallowest places in the lake and this is the only [other] place where any trouble occurred.

CAUSE

At the time these fish were killed in Upper Gar Lake the people in the vicinity immediately formed an opinion that a poisonous condition existed in the water and that the fish would be unfit for use. This was not the case.

Our investigation revealed that there was absolutely no dissolved oxygen in the water of Upper Gar Lake. Naturally, fish life would be destroyed under such conditions. The cause for this depletion of oxygen was the direct result of temperature conditions which in turn had its killing effect on the profuse growth of algae.

TEMPERATURE CYCLE OF WEST LAKE OKOBOJI
Kenneth Lang

Wetland and lake ecosystems combine to form a larger, standing water ecosystem. Filtered water from unimpacted wetlands flows cleanly into our large recreational lakes. To understand this larger system it is necessary not only to understand wetlands and how they change but also to understand our lakes and how they change. The wetland cycles that most concern us occur on the order of decades and revolve around water levels. The lake cycles that most concern us occur annually and center on water temperature.

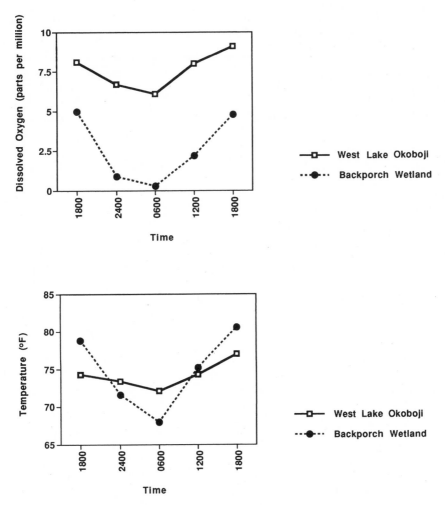

*Typical daily oxygen (top) and temperature (bottom) fluctuations of a
Gull Point wetland and nearby West Lake Okoboji. In the top graph note
the relatively low dissolved oxygen levels in the wetland, particularly at
night, compared with the lake. In the bottom graph note the much greater
temperature fluctuations (higher temperatures during the day, lower at night)
in the wetland compared with the lake. Lower wetland water volumes reduce
the buffering capacity of these basins.*

We are always aware of the seasonal cycles on the land around us. Our
winter clothes become more deeply buried in the closet as winter and spring
give way to summer. Likewise, we are again reaching toward the back closet
for our coats when the first frost reminds us of winter. The organisms of
the lake are also carrying out their life cycles in response to these seasonal

changes. Their environment, however, changes less than ours in some ways. The biggest difference is that the temperature extremes of summer and winter are not nearly so great. A fish in the middle of the lake may experience a summer high of 75° F and a winter low of 32° F; a pleasant prospect compared to life on the shores of Okoboji. This relative stability compared to the land comes from a property of water termed specific heat. Water has a higher specific heat than air. (Specific heat is a number defined as the amount of energy required to raise the temperature of a substance 1° C [about 2° F].) So compared to air, temperature changes in the lake are always slow to develop.

But the lake environment does have its seasons, and no aspect of the natural history of Okoboji has generated more confusion and false information than the lake's annual cycle. It works like this. The two engines of sun and wind are the main forces responsible for lake seasonality. When sunlight is absorbed by the water it is transformed into heat. In fact, about half of the light energy that strikes the surface of the lake is absorbed within the first three feet, and the rest becomes absorbed soon after. Therefore, if there were no wind to mix the water, the lake would have a layer of warm water at its surface while the deeper water would remain cold. This warm water would sit on top of the cold because it is less dense. We need to digress briefly.

Water is said to have an anomalous, or irregular, change in density as it heats up. Most substances are most dense when in their solid (frozen) state and become less dense as warmer liquids. But water reaches its maximum density as a liquid, at about 39° F, becoming less dense both as it cools toward freezing and less dense as it warms up. This is important to the lake for two reasons: (1) ice floats; and (2) the greater the difference in temperature between two levels in the lake, the greater the difference in density, and therefore the harder it is to mix the two layers. The way water mixes is probably the most important single factor to remember in understanding the seasonal dynamics of the lake.

Imagine West Lake Okoboji on a sunny, windless morning in April. By 10 o'clock the upper three feet will have warmed considerably, and if you are brave or crazy and slide into the lake, your toes will tell you where the cooler water starts. But most days aren't like this. Most days will be windy and even if there is just a breeze, the warm water will not form layers (stratify) at the surface.

When the wind blows across the surface of the lake a thin layer of water is moved in the direction of the wind. This is a surface current, and you can see it if you look carefully. This water travels to the far shore, where it is

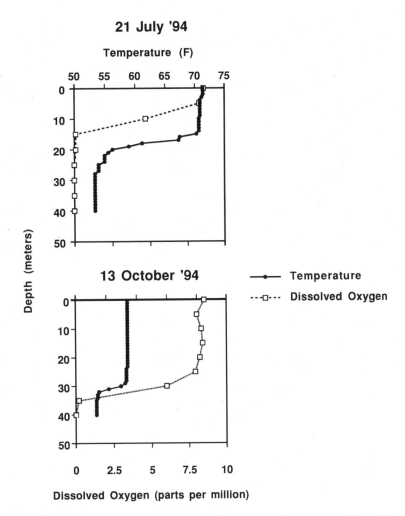

Summer (top) and autumn (bottom) temperature and oxygen profiles of West Lake Okoboji. Note that as the surface waters of the lake cool in autumn, the surface temperatures penetrate much deeper into the lake. Furthermore, high oxygen levels tend to follow the temperature profile.

deflected downward and sets up a return current traveling in the opposite direction, underneath the surface current. And on the near shore, another return current is set up under that, and so on. Where these currents contact each other, small eddies of turbulence are set up and water is mixed between the layers. In this way, the warmed water of the upper three feet is mixed with the cooler waters below. When the water column becomes the same temperature throughout, this mixing between layers can go on all the way to the bottom. Although not very well named, this slow mixing to the bottom is called "turnover." The name, of course, sounds more dramatic than it actually is. And although this mixing is going on throughout the lake in May and early June, you won't notice anything remarkable. But something important is going on that will come to affect the lake for the rest of the summer. This is where the density and temperature relationship of water becomes important.

As the lake is warming and being mixed by the wind, the temperature of the bottom waters lags behind the surface. This occurs because most of the time the wind is not strong enough to overcome all of the friction between each of the currents all the way to the bottom. So the bottom waters warm more slowly. Furthermore, the temperature difference with the surface becomes greater each day. This means that even more wind is required to mix the warming waters of the surface with those underneath. This process goes on until finally the warming of the upper layers gets so far ahead that a boundary layer, called a thermocline, develops between the upper mass of warm water and the lower mass of cold water. This thermocline in West Okoboji begins at about thirty-nine feet and extends downward to about fifty feet. Above the thermocline, the warmer water mixes. Below the thermocline, the water gradually cools with depth. In most years the temperature at the bottom is about 58° F.

Once the thermocline has developed, we say the lake is stratified (into a warm layer above the thermocline, the thermocline itself, and a cold layer below the thermocline). This is an important event in the seasonality of the lake because the thermocline now separates the surface water from the deeper water, and they can no longer mix. The oxygen of the surface water cannot get to the bottom, and the plant nutrients of the bottom waters cannot reach the surface. The tiny plants of the plankton starve for nutrients, and the animals confined to the deeper waters have little or no oxygen. During most years, stratification occurs by July.

Stratification usually persists well into September or even early October, when the lake cools down. But in reality, the lake begins to cool down in the latter half of August—less energy is being absorbed at the surface as the days get shorter and the sun becomes lower in the sky. When the amount of energy from sunlight being absorbed is less than the amount of heat energy radiating out, surface waters begin to cool and sink downward. As the upper body of water cools the thermocline recedes and mixing becomes easier. Finally, some windy autumn day, when the leaves are changing colors, the last fragment of the thermocline disappears and the lake once again mixes from the surface to the bottom—the (again inappropriately named) fall turnover.

By December, the lake will have cooled to 39° F from top to bottom. The water mass is at its highest density. Further cooling will lower the surface temperature toward freezing. Since this colder water, which you will recall is less dense, now floats on top of the warmer 39° F water, a winter stratification begins, which culminates with the freezing of the surface water. The lake is now sealed-off from the wind and little mixing occurs until the ice once again melts in March or April.

Notice something, that both in summer and in winter the lake is stratified by temperature (and therefore by water density). In both the summer and the winter the least dense water lies over the most dense water. In the summer, it is the warmest, least dense, water that comes to lie over the coldest, most dense, water. In the winter, it is the coldest, least dense, water that comes to lie over the warmest, most dense, water. In both summer and winter, the deepest waters are the most dense—the waters that approach 39° F. Because the summer surface waters are the warmest and the winter surface waters are the coldest, the spring and fall turnovers have been termed "flip flops." From this unfortunate terminology, some people have tended to think that they should avoid the lake during this time; that a fisherman in a boat at the surface at the time of a flip flop would somehow find himself upside down at the bottom of the lake when it was all over. This is, of course, ridiculous. Funny thing is, once people have this image in their mind, it becomes difficult to erase.

IMPACT OF SETTLEMENT ON WEST LAKE OKOBOJI
Richard Baker

A record of the environmental change of Okoboji is recorded in the mud of our lakes. Studies of the cores of the bottom sediments from Little Miller's Bay indicate that Euro-American settlement and the beginning of farming affected not only the prairies and the woodlands of the area but the lake itself. A large number of acres went into cultivation in the Okoboji area between 1863 and 1865. Trees at the time bordered both East and West Lake Okoboji. These trees, mostly oaks and hickories, were extensively cut for firewood between 1865 and 1880.

Plant remains preserved in the lake mud tell this story. A great increase in ragweed and goosefoot pollen marks the beginning of cultivation in Okoboji, indeed throughout the eastern United States. These plants are weeds that are specifically adapted to colonizing disturbed ground. On native prairies they grow on gopher and badger mounds; they thrive following plowing—the ultimate disturbance. Once the prairie sod is broken, not only does corn flourish, but weeds as well. The cutting of trees is also reflected by the decrease in oak and hickory pollen beginning at this same time. Furthermore, the planting of ashes and elms in towns and as farm wind breaks in the last few decades is recorded by increases in the pollen of these trees.

But how did deforestation affect the lake? There is a sharp change in the kind of sediment deposited in Little Miller's Bay at about fifteen inches deep into a core. Instead of the highly organic "mud" that is normally found some distance from shorelines, and that occurs below fifteen inches, a much siltier, sandier sediment abruptly appears. This sediment type reflects the substantially increased runoff that characterizes row crop agriculture. This newer sediment was deposited, and continues to be deposited, at a rate of about four-and-a-half times faster than sediment deposited before settlement.

Another change that is seen in the mud was apparently caused by the spillway built in 1910 that raised the level of the lake and regulated the lake level (although during the summer of 1993 the effectiveness of this strategy might be questioned by residents living in low-lying areas). At this time a rapid decline occurred in certain shallow-water aquatic plants such as the sago-pondweed, water-milfoil, and horned pondweed, whereas naiads—plants adapted to both shallow and deep water—continued at previous levels. The sudden deepening of the lake clearly affected the aquatic vegetation in the shallower parts of the lake. What else did it affect? At this point we do not know.

The messages read in the bottom cores of West Lake Okoboji are: even in such a large lake, human-caused changes, both large and small, have measurable and long-lasting effects that are often far-reaching and unpredictable; it is clear that we do not understand this complex aquatic ecosystem well enough to justify purposely changing it for our own short-term convenience.

4   HISTORICAL FAUNA

Okoboji is a transitional region. To the east lies the eastern deciduous forest, which extends fifteen hundred miles to the Atlantic Ocean. To the west the Great Plains extend five hundred miles to the foothills of the Rocky Mountains. To the north lies the poorly drained Minnesota lake region. To the south is the well-drained landscape of the Missouri and Mississippi River tributaries. The native animals of Okoboji reflect this confluence. For example, in the common names of our amphibians we find the eastern tiger salamander, the western chorus frog, the northern leopard frog, and the Great Plains toad.

To know how far Okoboji has come—how much we have changed—we must know the animals that were historically present and have some idea of their numbers. The essays by Stephens and Larrabee describe the historical vertebrate fauna of Okoboji. The articles by Schramm and an anonymous writer give us some idea of the historical abundance of Okoboji's animals. Mac Farland describes life as a student at Lakeside Lab in the earliest days.

ABOUT THE AUTHORS

Thomas Calderwood Stephens was trained as a physician but never practiced medicine. Instead, he became a member of the Morningside College faculty beginning in 1906 and stayed for forty years. Stephens was an influential teacher who taught at Lakeside for sixteen years off and on between 1911 and 1948. One of Stephens's Lakeside Lab students in 1911, Ira Noel Gabrielson, went on to become the first director of the U.S. Fish and Wildlife Service under Franklin Roosevelt. The T. C. Stephens State Forest, in southern Iowa, is named after this early Iowa naturalist.

Charles W. Schramm was an amateur naturalist, a resident of Des Moines who spent his summers in Okoboji. Schramm teaches us that one does not need to be a professional biologist to make a contribution to our knowledge of natural history.

Austin P. Larrabee is known for his 1926 monograph on the fishes of the Lake Okoboji region, from which the article included here is drawn. We do not know much about him. We do know that he worked at Lakeside during the summers of 1921, '22, '24, and '25, and that he worked closely with Professor Stromsten at Lakeside and the biologists at the Spirit Lake Hatchery.

Frances Mac Farland was a student at Lakeside Lab during its first years and recalls the days when the Lab was a private institution.

MAMMALS OF THE LAKE REGION OF IOWA
T. C. Stephens, 1922

"This country has been emphatically a game region, even since your correspondent came. There were once buffalo and elk all over these beautiful prairies. The buffalo were scarce, it is true, sixteen years ago [1866], but there were some. Of elk, there were large herds; but now they have passed away and are seen no more. This was the trapper's paradise. Otter, beaver, mink, badger, fox, wolf, and muskrat were here in great plenty; even now these animals are quite abundant in some seasons, except the beaver, of which a very few are caught each year, and of the otter none. Foxes are very plenty; also prairie wolves, and annually a few deer are shot. Rabbits (hares properly) are in great plenty, and occasionally a jack rabbit is seen. A few Canada lynxes are here, but they are rare; four have been killed here since I came. The fur-bearing animals bring quite a revenue yet, and they will continue plenty for a long time yet. There are no squirrels except the red squirrel and striped and gray ground squirrel. Gophers are plenty."

The above lines were written in 1882 by Mr. A. A. Mosher, of Spirit Lake, and published in *Forest and Stream* of that year. A few of the animals mentioned in this letter are still [1922] to be found in the lake region in decreasing numbers, while some that are now found sparingly are not mentioned by Mr. Mosher, such as the woodchuck and the raccoon. Of course, the species which interested Mr. Mosher were those which were classed as game or fur-bearers; and he omitted mention entirely of the smaller forms like mice, voles, shrews, bats, etc. But the omission of two such well-known and conspicuous animals as the prairie dog and the porcupine may justify the conclusion that they were not known in the locality. . . .

Virginia opossum. I do not know that the opossum has ever been found within the limits of Dickinson County. . . .

Virginia deer. In 1882 Mr. Mosher wrote that "annually a few deer are shot."

American elk. The elk is known to have covered this entire region in earlier years.

American buffalo. It formerly roamed all over the prairies.

Fox squirrel. This species is fairly common along the wooded borders of the lakes. . . . Mosher mentions the presence of the "red squirrel" in the [eighteen] eighties; if he thus referred to the Chickaree, which is probable, then it is likely that he was not aware of the presence of the fox squirrel. This raises the question as to whether one of these two squirrels has become recently established in our region.

Gray squirrel. . . . so far as I know it has never been found in Dickinson County.

Red squirrel. Chickaree. This saucy and vivacious little animal seems to be pretty well established in the wooded areas of the Lake region.

Gray chipmunk. The chipmunk seems to be very common in the wooded portions of the Lake region.

Striped (thirteen-lined) ground squirrel. This well-known rodent pest is very abundant throughout the Lake region. . . .

Gray (Franklin) ground squirrel. This gopher is also rather common, though it seems to recede more promptly before human occupation than does the preceding species.

An Okoboji trapper's shack. The smaller animals are mink; the larger is a coyote. Photo courtesy of Steve Kennedy and the Iowa Great Lakes Maritime Museum.

Woodchuck. . . . I would be inclined to consider the woodchuck fairly common in the wooded areas. In the summer of 1912 one of them had his abode under the cottage immediately south of the Lakeside Laboratory [its descendants still do].

Beaver. Doubtless exterminated at present.

Muskrat. Muskrats were plentiful in former years (Mosher); and even now they occur in considerable numbers throughout the county.

Pocket gopher. The pocket gopher is present in the Lake region, but apparently not in as great abundance as in other parts of the state.

White-tailed jack rabbit. Mosher reported the jack rabbit as "occasionally seen." They are still occasionally seen. Two were frequently seen in 1921 in the fields adjacent to the Lakeside Laboratory.

Cottontail rabbit. The rabbit, or hare, was considered by Mosher to be plentiful in numbers in the eighties. It is now to be found in relatively small numbers.

Mountain lion. The panther probably ranged over the entire prairie region in former times.

Bobcat. Mosher reports, "A few Canada lynxes are here, but they are rare;

An Okoboji trapper's haul at the turn of the century. Note the abundance of mink hides in the center portions of both rows of pelts. Photo courtesy of Steve Kennedy and the Iowa Great Lakes Maritime Museum.

four have been killed since I came." While the Canada lynx is here specified, the probability is that the distinction was not made between the two lynxes; and the bobcat, or wildcat, is the one most likely to have been found.

Red fox. This species is still occasionally found in our region.

Timber wolf. There is no satisfactory evidence that the real, big, gray, timber wolf has ever been taken or seen in this part of Iowa. Most of the reports of wolves refer to the coyote, or prairie wolf.

Coyote. Doubtless this animal was formerly abundant, but it is now becoming very rare.

Otter. Mosher's notes tell us that otters were plentiful in the eighties, but they have probably disappeared long since. Nothing is known of their occurrence in recent years.

Skunk. The true skunk is believed not to be very common in this part of the state.

Prairie spotted skunk. This is the only species of civet cat found in western Iowa, and it is more numerous than the skunk.

Badger. These animals have been much more common in the past but are gradually disappearing, since man never loses an opportunity to destroy one.

Mink. Reported by Mosher as plentiful in the eighties, and it is still found sparingly.

Weasel. The writer has not been able to examine specimens of the long-tailed weasel, which undoubtedly occurs in the region. . . .

Fisher. I have no definite information on the occurrence of this species in the lake region. But I have a vague recollection that someone reported to me orally the capture of a specimen in earlier years. More than likely it oc-curred here formerly in small numbers at least.

Raccoon. Mr. Harry C. Tennant has a mounted coon which was taken in 1916. Mr. Tennant considers the coon to be "fairly common" in numbers.

Black bear. A news item published in the Sioux City *Journal* of Novem-ber 29, 1876, says that "a large black bear was seen in the vicinity of Spirit Lake a few days ago."

THE BIRDS OF THE LAKE REGION
T. C. Stephens, 1918

Many summer visitors at the lakes will be surprised to learn that there are about 125 different kinds of birds around the lakes during the sum-mer months. For instance, . . . western meadow lark, prairie horned lark, dickcissel, killdeer, kingbird, flicker, bobolink, short-billed marsh wren, vesper and grasshopper sparrows, upland sandpiper, warbling vireo, brown thrasher, red-headed woodpecker, yellow warbler, barn swallow, belted kingfisher, spotted sandpiper, black terns and Forster's terns, bank swal-lows, rough-winged swallows, house wren, bluebird, rose-breasted gros-beak, Baltimore oriole, catbird, bluejay, bronzed grackle, robin, phoebe, wood pewee, scarlet tanager, indigo bunting, red-eyed vireo, crested fly-catcher, song sparrow, chickadees, yellow-billed cuckoo, Cooper's hawk, prothonotary warbler, wood thrush, towhee, goldfinches, coots, pied-billed grebe, common gallinule, blue-winged teal, bluebill, lesser scaup duck, mal-lard, red-winged and yellow-headed blackbirds, prairie marsh wren, bittern, black-crowned night heron, great blue heron, swamp sparrow, Maryland yellow-throat, king rail, Virginia rail, sora rail, nighthawk, tree swallow, cliff swallow, yellow-throated vireo, purple martins, and chimney swifts.

CROWS
Anonymous, 1922

Attacked by thousands of crows while duck hunting on the Missouri River near Sioux City, M. L. Murray of Salix killed and wounded more than five hundred of the angry birds in three hours of constant fighting, using about two hundred shotgun shells. The coming of darkness gave him an opportunity to escape. The fight started when Mr. Murray shot at a lone crow and wounded it. The wounded bird cried for help and soon hundreds of others came to its aid. Soon the ground was covered with dead and wounded crows. The remaining birds attacked Murray, striking him with their beaks and wings. When darkness came the hunter was able to retreat to a willow patch and the crows, unable to advance upon him, returned to roost.

[In light of this story, it is ironic to learn that a large gathering of crows is termed a "murder."]

FISHING IN 1884
Chas. W. Schramm, 1914

Of course we fished. What Fishing! We did not trawl for pike in those days but anchored our boats and caught them still-fishing, in great quantities. Without an attempt at irresistible levity or wishing to indulge in a great amount of sarcasm, I feel justified in making the statement that fish in those days behaved in an altogether different manner from that now attributed them, particularly in their habits of diet. I have seen tens of thousands of buffalo and gar fish sunning themselves on the water's surface, while the pike would strike with such regularity. . . .

THE FISHES OF THE OKOBOJI LAKES
Austin P. Larrabee, 1927

The fish life of the Okoboji lakes furnishes one of their most attractive features. . . . The number of fishes found in any body of water depends upon a variety of factors. Among those that are most evident may be mentioned the oxygen content of the water, the food supply, the type of bottom, shelter, and suitable places for spawning. Forty-five species in all have been recorded from the Okoboji lakes of which about forty are native to them. This number indicates that conditions here are favorable for fish life. The variety of habitats afforded by the lakes undoubtedly explains the large number found. For the lakes furnish different kinds of bottom, muddy, sandy, and rocky, varying amounts of vegetation, and varying depths, especially in West Okoboji Lake. Certain species such as the carp and bullheads prefer a muddy bottom. Others as the shiner or spot-tailed minnow and the sand darters show a preference for sand, while still others as the wall-eyed pike are found most commonly where the bottom is rocky. The muddy bottom attracts the greatest number of species due probably to the greater supply of food to be found there. Of thirty-eight species observed by the writer, twelve showed a preference for muddy bottom, nine for sandy, and two for rocky. Other species were taken under different conditions. Thus nine species were recorded over both mud and sand, four over both muddy and rocky bottoms, one over sandy and rocky, and still another on all three types of bottom. The aquatic vegetation similarly attracts the fishes, for twenty-six [species] were noted here, nine practically all the time and the remaining seventeen part of the time. Again the food supply furnished here is doubtless an important fact together with the shelter afforded by the plants. Twelve species showed a marked preference for open water. The carp, golden shiner, pickerel, and the common sunfish are characteristic fishes of the weedy portions of the lake. The straw-colored minnow, the banded killifish, and the darters are found most commonly in the open water between the weeds and the shore, while the wall-eyed pike is a typical fish of the deeper waters beyond the weeds. The perch is the most cosmopolitan of all and is found at various depths and under varied conditions. The association of fishes with water plants is illustrated by a comparison of some of the lakes of this region. Welch Lake with very little growth of the higher aquatic plants had but three species of fishes recorded from it, while Spirit, West Okoboji, and East Okoboji with relatively increasing amounts of water weeds had thirty-three, thirty-six, and forty species listed from them respectively. An apparent exception to this is found in Robinson Lake [the southwest basin of Hottes Lake] with a relatively greater amount of aquatic plants still, and with but six species recorded from it. The acidity of the lake as indicated by tests made of the water and the high temperature of the water in

Gigging for shortnose gar in Okoboji. Photo courtesy of Steve Kennedy and the Iowa Great Lakes Maritime Museum.

the summer may be factors accounting for this exception [as would be the size of the lake, the depth of the lake and the amount of nighttime and winter dissolved oxygen it could hold].

Of the forty species noted, there are two which may no longer occur. One of these is the spoonbill or paddlefish, an odd-appearing fish with a thin paddlelike snout overhanging the mouth, giving it an appearance somewhat like a shark. The use of this paddlelike organ is not known, although it is commonly regarded as sensory. [We now know that the paddle is studded with thousands of electroreceptors.] Three very large specimens over six feet in length have been taken in years past, all during the winter. The largest one, a male taken in the winter of 1915–1916, measured six feet and nine inches and weighed 185 pounds. This is the largest specimen of this species yet recorded, exceeding the one taken in Lake Manitou, Indiana, which weighed 173 pounds. The spoonbill is found more commonly in rivers and larger streams, preferring the quieter waters of these where the bottom is muddy. The large size attained in the lakes may be due to a considerable degree to the abundant supply of the small crustaceans and the insect larvae which constitute the bulk of its food. No spoonbills have been taken since 1918, although fishermen have reported seeing others.

The second species which is doubtfully present in the lakes is the Great Lake trout, an introduced species. At different times during the past twenty

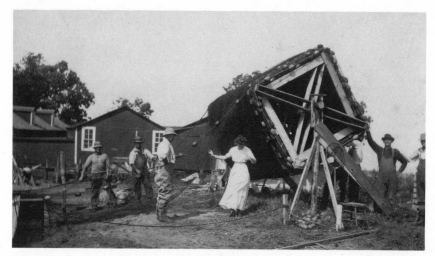

Photographic postcard showing the big Okoboji commercial fishing nets being rolled up for drying and storage. Photo courtesy of Steve Kennedy and the Iowa Great Lakes Maritime Museum.

years plantings of this fish have been made in Spirit and West Okoboji Lakes. In 1922, 40,000 adults were introduced into the latter lake. The results of these attempts have not been satisfactory so far. About fifteen years ago, four or five trout were caught in the lakes. One of these, taken in Spirit Lake, was mounted, and in the summer of 1925 was still hanging on the wall of a pool room in the town of Spirit Lake. West Okoboji Lake evidently affords more nearly the conditions required by this trout, because of its depth and the cool water found there in the summer. Apparently the amount of oxygen in the deeper waters of the lake becomes too much reduced in the summer for this species, although it can endure a much smaller amount than the average fish.

Another fish belonging to the same order as the trout [this view is no longer held by ichthyologists] but to another family, the northern mooneye, occurs rarely in the lake. Two instances of the capture of this fish have come to the author's notice. One of these was taken in East Okoboji in the summer of 1922, the other in Minnewashta Lake some years ago.

Among the common species of fishes found . . . in the weedy portions of the lakes are the two species of gar-pikes, the long-nosed gar and the short-nosed gar. During the summer the latter is seen more commonly than the former. Both have the habit of floating like a stick on the surface of the water. Their bodies are nearly cylindrical and covered with hard rhombic

Photographic postcard showing large hoop nets. These nets can be used with or without bait and are an efficient way to trap turtles. Photo courtesy of Steve Kennedy and the Iowa Great Lakes Maritime Museum.

scales. Their elongated jaws armed with sharp teeth suggest those of an alligator or crocodile. They are a voracious fish, the food of the short-nosed, according to the observations of Mr. G. E. Potter, consisting of about 60 percent fish and 40 percent crayfish. Probably no fish, with the possible exception of the carp, is more thoroughly disliked by fishermen than are these two species.

During the summer season the members of the sucker family are not much in evidence. Two species of buffaloes occur, the large-mouth buffalo and the small-mouth. The former is the much more abundant of the two. In the summer both prefer weedy waters six to eight meters deep. The common sucker is another well-known representative of this family, occurring commonly in the three major lakes and in the Gar Lakes. Great numbers of these suckers are taken in the seining operations during the fall. The red-horse has been taken in East and West Okoboji Lakes but is evidently an uncommon fish here.

The minnow family, the largest one among fishes, includes the greatest number of species of any family represented in these lakes. Eleven species of this family have been recorded. To this family belong the fishes commonly termed minnows, chubs, shiners, and daces. They are much used as food by the larger predaceous fishes and thus form an important link in the food cycle of the lake, and in determining the number of game fishes to be found

Photographic postcard showing an Okoboji commercial fish-holding pen. It is dated January 1906. Photo courtesy of Steve Kennedy and the Iowa Great Lakes Maritime Museum.

there. While generally small and inconspicuous, some notable exceptions occur. The introduced carp attains considerable size, and the silver chub together with the golden shiner are moderate sized fishes. The carp, a fish thoroughly disliked by the fishermen, has become very common in the lakes since its appearance about 15 or 16 years ago. It is commonly accused of eating the spawn of other fishes, although the evidence against it is not wholly conclusive, of rooting up the vegetation, of muddying the water, and of driving other fishes away. In its support it has been suggested that the young carp may serve as food for the more valuable game fishes. It is found commonly in the weedy portions of the lakes. In these places, too, may be found the golden shiner. This fish evidently prefers the shallower lakes, although it is found in lesser numbers in the deeper lakes. The silver chub is reported as common in the Little Sioux River, and as occurring occasionally during the spring run in the lakes.

Of the smaller members of this family, the spot-tailed shiner is the most common. At one time this was the common minnow used for live bait, but its numbers have been reduced so that it has been supplanted largely by the blunt-nosed minnow and the bullhead minnow. The last two occur rarely

Hauling seines across the ice for winter commercial fishing in Okoboji. Photo courtesy of Steve Kennedy and the Iowa Great Lakes Maritime Museum.

in the three larger lakes but are very common in some of the smaller ones. The bullhead minnow was one of the three species recorded from Welch Lake. Five other minnows occur in lesser numbers. The blacknose shiner and the blackchin shiner are rather uncommon here, the former being found tolerably common in Sunken Lake, the latter in fair numbers in a kettle hole near the shore of East Okoboji. The sand shiner is another of these small species that occurs sparingly in the lakes. This one is common in the Little Sioux River and was found in more limited numbers in two small creeks on the east side of East Okoboji. Another minnow that is more strictly a river species is the red shiner. This is found commonly in the Little Sioux, and but two specimens were taken in the lakes. It seems possible that these were individuals which had escaped from bait boxes. The male of this species is a very pretty little fish with a bright blue body and red fins and is an attractive fish for an aquarium. The river shiner appears to occur mainly in the region of Gull Point, West Okoboji Lake, and is [locally] called the Gull Point chub for that reason. It is a rare species evidently, for the author has taken but one specimen in his seining operations.

The catfishes and bullheads are familiar fishes, easily recognized by the absence of scales on their bodies and the prominent barbels near the mouth. Two species of the larger catfishes have been noted, the channel cat and the flathead cat. The former is probably a native here, as specimens have been caught in the Little Sioux River. Its numbers have been increased in the lakes by the State Fish and Game Department introducing them there. This stocking of the lakes was first begun about thirteen or fourteen years ago

In Okoboji, fishes (and frogs) were stored in snowdrifts until an outbound train took them to market. Photo courtesy of Steve Kennedy and the Iowa Great Lakes Maritime Museum.

and was evidently successful. Young catfishes were first noted in the summer of 1921. The occurrence of the flathead is apparently accidental, some of them being introduced with the channel cat. Three species of bullheads are found in the lakes. The yellow bullhead, also called the yellow cat, is more common in East and West Okoboji, while the black bullhead is the more common in Spirit Lake.

The common pike or northern pickerel is a fine large gamefish, well known to the sportsmen. It is an active predaceous fish, preferring the neighborhood of the water plants, where it lies in wait for its prey, mainly fish, capturing it with a sudden dash. It is well fitted for its life, the rounded body making it easy for the pike to move in and out among the water plants. Its mouth is large and armed with strong sharp teeth for holding its victim. The fish when lying in wait resembles a submerged stick, which is doubtless of service in securing its prey.

Another fish belonging to the same order as the pike [ichthyologists no longer classify these fishes in this way] but to another family is the banded killifish. This is a small fish found generally in the shallow water between the weeds and the shore or in the weeds themselves, showing a preference for a sandy bottom. Although an attractive looking fish and not uncommon in some parts of the lakes especially in late summer, it generally escapes notice or is simply called a minnow. For the term minnow is used to include not only true minnows but also any small fish and even the young of larger fishes.

The trout perch is a peculiar fish, combining some of the characters of the salmons and perches. [The ichthyologist] Dr. [David Starr] Jordan, speaking of the family to which this fish belongs, says "it would seem to find its place in Cretaceous rocks rather than in the waters of today." . . . Dr. B. W. Evermann, who visited the lakes in 1892, states that this fish was considered abundant at the time. At the present time, however, it is uncommon. It has a somewhat translucent body and suggests the darters in general appearance and actions. The author took but one specimen of this species in the lakes, that one from Spirit Lake. Others were taken in the Little Sioux.

The family to which the crappies, sunfishes, and bass belong is represented in the lakes by nine species. Most of these are of value as food fishes and the basses are two of the most important freshwater gamefishes. The fishes of this family have the body more or less deep and flattened from side to side. They are generally brightly colored, sometimes strikingly so, especially the sunfishes. Although the coloration is apparently conspicuous, it is in reality protective or aggressive, tending to blend with the lights and shadows of their habitat, which is typically among the water plants. Their flattened body enables them to move with little hindrance among the stems of these plants. Two species of crappies are found in the lakes, the white crappie and the black crappie or calico bass. The two are similar in habits and distribution, showing a preference for waters of moderate depth among the water weeds and coming into the shallower water in the evening for feeding. Judging from the results of seining, the white crappie is the more common of the two. . . . The rock bass is found in all three of the major lakes but is not common in any one of them. Because of its prominent red eyes, it is sometimes called the "red-eye" and also the "goggle-eye." It is only a moderate-sized fish but is a very good food fish.

Four species of sunfishes occur in the lakes. The most common species and the largest one of the four is the bluegill. With the exception of the perch, the bluegill is the most abundant fish in the lakes and is found under a variety of conditions of depth and bottom. The common sunfish, or pumpkin-seed, is somewhat more widely distributed than the bluegill, occurring in the shallower lakes as well as in the three major ones. It is a

One of the most frequently reproduced of all historic Okoboji photographs.
Big paddlefish were commonly taken in West Lake. During the winter
of 1915–1916 three fish, each over six feet long, were captured. The largest,
at six feet, nine inches and 185 pounds, was at that time the world's record.
Is West Okoboji richer for no longer having these animals? Photo courtesy
of Steve Kennedy and the Iowa Great Lakes Maritime Museum.

strikingly colored fish and is well known to all fishermen. Associated with it
in the shallow lakes, Hottes and Robinson [today both of the lakes are
called Hottes; the southwest basin was formerly called Robinson], in the
summer of 1922, was found the long-eared sunfish. This species is even
more brightly colored than the common sunfish and has an elongated flap
to its operculum or gill cover. Apparently it occurs rarely in the major lakes,
for the author has known of but two specimens taken from them, both of
these in East Okoboji. It doubtless is found in Spirit Lake and probably in
West Okoboji as well. The remaining sunfish, the green sunfish, is a common fish of the weedy bays of West Okoboji Lake and occurs in the other

One of the large Okoboji paddlefish, well over six feet in length. Note that it reaches from the roof trim to the deck of this porch. Photo courtesy of Steve Kennedy and the Iowa Great Lakes Maritime Museum.

larger lakes in lesser numbers. All of the sunfish prefer the weedy portions of the lakes, and make their nests among the plants in the shallow water near the shore. These nests are guarded vigorously and with great courage by one of the pair, generally the male. The nesting is mainly in the early summer but may continue into August.

The large-mouthed bass is one of the favorite game fishes of the lakes and occurs tolerably common in the three major lakes. Occasionally the small-mouthed bass is taken also. This species has probably been introduced accidentally with the other, although Professor S. E. Meek, who visited this region in 1890, mentions taking this species in Spirit Lake.

The perch family is represented by six species. Three of these are small fishes commonly known as darters, of which the most common one is the

A day's catch in turn-of-the-century West Okoboji. This photograph was taken in Arnold's Park. Can you identify the fish? Photo courtesy of Steve Kennedy and the Iowa Great Lakes Maritime Museum.

sand darter or log-perch. Another one, the Johnny darter, easily confused with the former, is more common in the Little Sioux than in the lakes. These two are inconspicuously colored and match well the color of the sand, their usual habitat. The third darter, the Iowa darter, is more brightly colored than the other two but resembles them in its habits and habitat. All three have the habit of resting on the sand, darting suddenly when disturbed and thus getting their name of darters.

The other three species of the perch family are much larger and include the perch, the wall-eyed pike, and the sauger. The perch is the most common as well as the most widely distributed fish of the lakes, occurring under varied conditions of depth and of types of bottom. It is an active predaceous fish, a fine example of an "all round fish." The largest perch are taken in Spirit Lake. The wall-eyed pike is the most important game fish of the lakes and is found in considerable numbers in all three major lakes. . . . The related species, the sauger, is said to occur in all three lakes with the wall-eyed but in much lesser numbers.

Two species only remain [to be discussed] of the forty-five recorded by the author. One of these, the silver or white bass, is found commonly in East and West Okoboji Lakes and in somewhat lesser numbers in Spirit Lake.

The remaining species, the sheepshead, or fresh-water drum, is found commonly in East and West Okoboji Lakes but is comparatively rare in Spirit Lake. While it is sometimes eaten, it does not rank high as a food fish. The name of fresh-water drum or grunter is given to it because of the curious sound made by the fish, audible for some distance.

MISSING PIECES 1

From the early accounts of the fauna of Okoboji we can assess the vertebrates lost and gained. The most conspicuous losses are the large mammals. We clearly no longer have black bears, buffalo, elk, bobcats, otters, cougars, prairie spotted skunks, Franklin ground squirrels, and fishers. Deer, raccoon, and red fox numbers have increased; gray fox, striped skunks, and opossums are now here. Beavers, if they ever were completely gone, have become reestablished.

Conspicuous losses also occur among the birds. Scarlet tanagers and prairie-chickens are now missing; upland sandpipers are rare. Trumpeter swans were reintroduced in the spring of 1995. We have added cardinals, ring-necked pheasants from China, and gray partridge, starlings, and house sparrows from Europe.

Of Okoboji's native fish, the trout perch, paddlefish, northern mooneye, Storer's chub, and several minnows, including the blacknose shiner, are among those now locally extinct (as we will see, any fish population, such as the blacknose shiner, historically present in Sunken Lake was doomed). Lake trout and flathead catfish were introduced and are now gone. Introduced carp now thrive. Orangespotted sunfish, creek chubs, fathead minnows, and two catfishes—the stonecat and the tadpole madtom—are present. The muskie is an introduced game species that is being artificially maintained. Although not mentioned above, the prairie rattlesnake was here and is now gone (Pammel 1929).

These losses are stunning. Not only have we lost our large mammals—expected during the transition to an agricultural landscape—but we have eliminated many smaller species. Why do we have thirteen-lined ground squirrels and chipmunks but not Franklin ground squirrels? Why do we have striped skunks but not prairie spotted skunks? What caused the scarlet tanagers, prairie-chickens, trout perch, paddlefish, northern mooneye, and the several minnow species to disappear? If they returned, would modern Okoboji accommodate them, or have conditions changed—habitats been altered—too much? Are these species the temporary or permanent victims of Okoboji's progress?

In addition to losing species, it is clear that within the species that remain populations have been reduced. It is nearly impossible to conceive of the presence of so many crows that an armed man could not escape them. In fact, how many modern hunters expect to see so much game—even those who are lousy shots—that they now carry eight boxes of shotgun shells into the field with them? And who now can imagine seeing tens of thousands of suckers and gars on the surface of our lakes? So many fish were here that fishermen say they could watch the northerns feed on them. Who among today's fishermen has ever seen a northern pike feed, except for when it has been staged for a television program? I refer those interested in reading further about Iowa's historic game fauna to James Dinsmore's excellent book, *A Country So Full of Game.*

IOWA LAKESIDE LABORATORY
Frances Mac Farland, 1911

On a hill at the head of a quiet bay, overlooking a long vista of lake and shoreline, you will find a few buildings and a large flock of tents. These furnish headquarters for the people who are having the best time of anyone on the lake. A large sign announces "Lakeside Laboratory" to inquiring strangers on steamboats and other craft. If one lands here at the right time of a day and climbs the stairs to the top of the bluff, one finds the laboratories occupied by busy students working over live plants or animals which have been recently dredged up from the bottom of the lake or some neighboring bog, or collected from the woods along the shores. Other groups may be found listening to lectures out under the trees; others on the shore getting acquainted with the interesting and significant glacial boulders found here so abundantly. Later in the day students and instructors scatter

Students during the first years of Lakeside embarking on a field trip in an appropriately named boat. Again, note the formal attire of the time, even for field work. Photo courtesy of the Iowa Lakeside Laboratory.

toward all points of the compass. Some go by launch to explore distant parts of the lake, to dredge the bottom, and skim the top; some take to the deep woods; others travel "afoot and lighthearted," careless of sun and wind, over the prairie hilltops. More ambitious excursions go by train and by carriage to all natural objects for many miles around. To a student of natural science these are ideal conditions—conditions under which work is pure pleasure. In few laboratories can live material be obtained for study, and in few localities can the student do his work in the field studying plants and animals in their natural environment. But the individual student is not the only one who is benefited. Here is a place for pioneer work—original research—the results of which may add to the sum of knowledge and possibly be of great importance to all the people of Iowa. Every year teachers from the various schools of the State become students in this laboratory, and carry back to their own teaching fresh enthusiasm, and a broader working knowledge, which cannot fail to raise the standards of science teaching in Iowa.

So wide-spreading are the benefits of the work done here that the Lakeside Laboratory properly belongs among the educational institutions supported by State appropriation. As a matter of fact, however, it is a private enterprise, supported and carried on by a corporation consisting of alumni of the State University. Such an arrangement does not guarantee the permanence and stability which should belong to work of this character, and it is to be hoped that the State Legislature will soon see fit to recognize this fact in some substantial manner. [This, in fact, has happened. Lakeside Laboratory is now a Board of Regents institution.] Meanwhile the public continues to benefit by the devotion of a few broad-minded, big-hearted men; and incidentally, Miller's Bay has become the center of a most interesting and delightful kind of college life.

Thousands of individual duckweeds, tiny vascular plants, cover the surface of the water in this portion of the west side of Little Spirit Lake. Muskrat-cut cattails provide perspective.

A classic Okoboji scene. The west side of Little Spirit Lake, ringed with cattails, near a bur oak savanna upland. Bur oaks are resistant to prairie fires and are found where the Great Plains grade into the eastern deciduous forest.

A western chorus frog adult, Okoboji's smallest amphibian.

A northern leopard frog, one of Okoboji's historic game animals.

An American toad, one of Okoboji's most often seen amphibians.

A tiger salamander adult, seen most often after warm spring and summer rains.

On gossamer wings. A dragonfly, the twelve-spot skimmer, in an Iowa wetland.

The colorful Michigan lily is found at the Freda Hafner Kettlehole.

A male northern shoveler, a duck with colors that can cause it to be mistaken for the more common mallard. But notice the spatulate bill on this bird, which it uses in a way that generated its common name.

The yellow flowers of the bladderwort are a common sight in Okoboji wetlands after the middle of June, when these plants have had the chance to grow up to the water's surface. This plant is Okoboji's only carnivorous species; it uses its bladders to feed on small aquatic crustaceans.

Painted turtles are found in almost every Okoboji wetland.

A male least bittern attempting to hide within the emergent vegetation of an Iowa wetland.

Yellow-headed blackbirds are among the most conspicuous animals in Okoboji wetlands.

A summer evening thunderhead, looking southeast over Hottes Lake.

One of Okoboji's most unusual wetlands is the fen that borders Silver Lake.

Cayler Prairie in full color on a late summer evening.

5 OKOBOJI'S WETLAND FAUNA

Wetlands are odd aquatic ecosystems. Aquatic animals that live in wetlands must be adapted to the extremes of the wetland environment—anoxia and drought. As a group, crustaceans—zooplankton, crayfish—insects, and mollusks—snails and clams—have the most numerous wetland representatives. Many of the birds that we associate with wetlands—red-winged and yellow-headed blackbirds and, of course, the various duck species—make use of this invertebrate prey base.

Amphibians are the vertebrate group most dependent on natural wetlands. Okoboji's seven species of amphibians must breed and raise their larvae in wetlands, and therefore can tolerate aquatic anoxic (no oxygen) conditions. As I will show, there are special features of amphibian larvae that allow them to convert the natural productivity of wetlands into their own productivity—into energy that they then make available in the form of prey for other, more familiar animals.

ABOUT THE AUTHORS

Robert Cruden and Richard V. Bovbjerg are both past directors of the Lakeside Lab and are both good friends. Cruden is in the botany department at the University of Iowa and is one of the most broad-based biologists I have ever met (what is a guy in a botany department doing writing about dragonflies?).

Bovbjerg is a Distinguished Iowa Scientist and was a professor at Lakeside for more than thirty years. Dick is now retired from the University of Iowa's zoology department. We at Lakeside owe so much to him.

Harry M. Kelley was a parasitologist at Lakeside and taught at Cornell College in Mount Vernon, Iowa. He was a member of the Iowa Academy of Science from 1894 to 1936 and was its president during 1915–1916.

Maude Brown was a student during Lakeside's second year and provides further details of Lab life at that time.

BLACKBIRDS

The most conspicuous animals in any wetland are the birds, and the most conspicuous birds of the emergent zone are the red-winged and yellow-headed blackbirds. Watch and notice how the red-wings nest to the outside of the marsh, while yellow-heads occupy more central territories. It is not always this way; this zonation develops every summer as follows. Red-wings migrate north in the spring earlier than yellow-heads and arrive first on Iowa wetlands. Red-wing males begin establishing and defending territories throughout the wetland basin. The yellow-heads then arrive. Yellow-headed males are larger and more aggressive than red-wing males, and drive them from their prime territory sites, taking over the center of the wetland. These territories are nearer the open water, where food—in the form of emerging insects such as dragonflies, damselflies, and caddis flies—is abundant, and protection from terrestrial predators is greater. But no need to feel sorry for the red-wings. They persist along the fringes of wetlands that contain yellow-heads and are found throughout wetlands that do not contain yellow-heads. In fact, red-winged blackbirds are not only our most common blackbird, they may be our most common marsh bird.

AQUATIC INSECTS OF OKOBOJI
Robert Cruden

Is there any nighttime visitor to Okoboji who has not stared with either amazement or horror at a brightly lighted window or screen alive with hundreds or even thousands of swarming mayflies, midges, and/or caddis flies? These legions are but a mere handful of those that started life as larvae the previous summer. As larvae, their numerous siblings were eaten by fish and larval amphibians. In fact, it is the form and behavior of these insects that fly fishermen strive to reproduce. As newly transformed adults, they become prey for adult frogs, salamanders, and birds, including swallows and nighthawks.

Mostly unnoticed as they emerge from lakes and wetlands are the dragonflies and damselflies, which are voracious carnivores in their own right. As aquatic larvae they stalk the smaller aquatic invertebrates. As adults, they become mosquito hawks, able to devour large numbers of our summer pests. For this behavior alone they deserve special consideration.

Imagine a dragonfly with a wingspan of eighteen inches patrolling a prehistoric wetland the way that modern dragonflies patrol the edge of the cattails at Jemmerson Slough. Have no fear, those monster odonates became extinct one hundred and fifty million years ago and were replaced by our current smaller, but no less voracious, versions.

Damselflies (suborder Zygoptera) and dragonflies (suborder Anisoptera) together form the insect order Odonata. Both groups inhabit the lakes, wetlands, and streams of Okoboji. Seven of the eleven families found in the United States are represented in the local fauna. Damselflies are represented by the black lacewings (family Agrionidae), spread-winged damselflies (family Lestidae), and narrow-winged damselflies (family Coenagrionidae). Dragonflies are represented by the clubtails (family Gomphidae), darners (family Aeschnidae), green-eyed skimmers (family Cordulidae), and the common skimmers (family Libellulidae).

Our odonate fauna has relatively few species compared to tropical regions and to areas of North America to our east and west. Yet the Okoboji fauna is exciting for its diversity. Like many Okoboji animals, odonate diversity includes eastern species on the western edge of their range, western species on the eastern edge of their range, and northern species on the southern edge of their range.

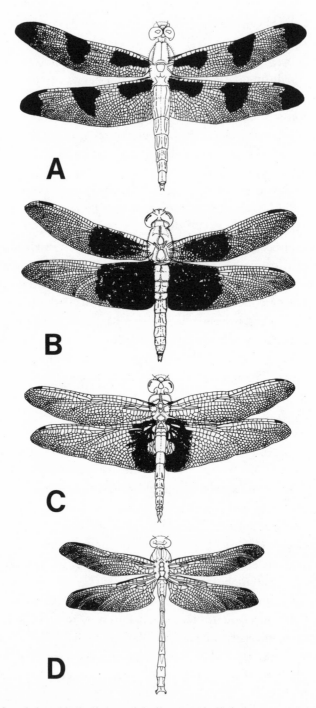

Dragonfly adults: A. Libellula pulchella; B. Libellula luctosa; C. Tramea lacerata; D. Gomphus graslinellus.

Wetlands are characterized by droughts, and it is fair to ask where odonates go during droughts. Do the few permanent ponds serve as refugia? The wet summer of 1993 was just three summers removed from a drought that dried all our temporary wetlands, yet we found large numbers of species. Were these abundant animals descendants from the few remaining individuals inhabiting the larger wetlands? Probably not, for we are finding species that are not normally associated with larger wetlands. Might the larvae of many species burrow into the bottom mud to wait for the return of water? This is possible and will be tested during the next drought cycle.

In 1909, the first year that Lakeside Lab opened, Arthur Whedon, a University of Iowa graduate student, collected dragonflies and damselflies along Miller's Bay and elsewhere in Okoboji. In 1993 and 1994, my colleagues, Kristina Baker and O. J. "Bud" Gode, Jr., and I repeated Whedon's survey. Happily, most of Whedon's species still occur in the region. In fact, the number of species now known from the region has more than doubled since Whedon's study. This increase is due to more extensive collecting and to species moving into the region during the past eighty years.

The bad news is that four species collected by Whedon, one from Spirit Lake and three from Miller's Bay, have not been recollected, despite two years of intensive searching. We collected one of these species elsewhere in northwestern Iowa. The other three have either been extirpated or are extremely rare in the region.

A second conclusion of our 1993–1994 survey is that while the odonate fauna of the wetlands and lakes has not changed too much in the past eighty-five years, the stream fauna has been altered. Many channeled streams, such as Pillsbury Creek, have either lost their odonate fauna or have seen it reduced. Where odonates occur, sometimes only one species, the black lacewing, is found. While Whedon found about twelve odonate species in Mill Creek (Cherokee County), including an array of clubtails, darners, lacewings, and narrow-winged damselflies, today it is a rare stream that supports as many as six species. As demonstrated by the sediments in West Lake, agricultural practices and stream ditching programs promote soil erosion at a level not seen when Iowa was covered with prairie and its streams meandered. As a consequence, silt smothers the bottoms of today's streams and restricts the plant growth necessary for odonate larval diversity.

AMBER LAND SNAILS

A number of small land snail species occupies Okoboji. One is found along the margins of our wetlands and is described in Lannoo and Bovbjerg (1985). The amber snail, *Oxyloma retusa*, a mollusk in the family Succineidae, is found only where it is damp. These snails are not usually found *in* water, but *near* water—typically not more than a few feet away from open water. If we were to map the distributions of the populations of this amber snail, they would either look like donuts ringing our wetlands or parallel, snaking lines along the banks of the Little Sioux River.

Why these unusual distributions? It appears that amber snails require high levels of humidity—at or near one hundred percent. Lower humidity levels cause them to lose water to evaporation and to begin to dry. With enough water loss they quit moving, form a mucouslike plug over the opening of their shell, and become dormant, a process called estivation. This is a temporary solution. If they do not become hydrated within a few days, they will die. Therefore, the ideal habitat for these snails is close to open water (and therefore high humidity), but not too close; they have a lung and can drown.

While called amber snails, these snails can range in color from a white-yellow wheat color to nearly black. The darker snails are found on the mudflats nearest the water; the lighter snails are found in the dead vegetation—detritus—a little farther upland. Like chameleons, individual snails have the ability to change color to match their environment. Bring the lightest and darkest colored snails into the laboratory and in a few days they will become the same intermediate color.

What happens to these snails during summers when wetlands begin to dry? As the water in the wetlands evaporates, water levels become lower, and the water margin recedes toward the deepest part of the basin. The amber snails follow the water margin and move toward the center of the wetland. But then what happens if it rains and the wetland refills? Amber snails will drown if forced to stay submerged. If their wetland refills, snails crawl up the stems of nearby vegetation. They then pass from stem to stem by voluntarily entering the water and crawling upside down on the water surface. Here, the surface tension forms a substrate that, improbably, allows these snails to attach and move. Eventually, the snails reach land or become food for wetland birds.

Amber snails usually avoid bird predation through camouflage—the reason for the color change. Unfortunately for these snails, they are the host to a parasite that has birds as its second host. For this parasite to complete its life cycle, a bird must eat its host snail. To make its snail more likely to be eaten, the parasite in its larval stage is colored red or orange. Larvae migrate to the snail's tentacle and pulsate up and down, flashing red—they are called pumpers. Infected snails stand out from non-infected snails the way police cars stand out from ordinary cars on a busy highway.

Amber snails overwinter by hibernating in the wetland plant detritus. Snails begin to hibernate after the first few frosts. By picking carefully through the detritus you can find hibernating snails. Collect some and keep them in your freezer during the winter. In the spring thaw them out, add water, and they will revive. If you thaw them out but do not add water, they never recover and will soon die.

THE CLAMS OF THE OKOBOJI LAKES
Harry M. Kelley, 1926

The casual visitor to the Okoboji lakes usually gains some acquaintance with the fish of the region, sometimes also with its birds and turtles, but only if curiosity prompts rather extended "putterin' round" its shores will contact be possible with the lesser aquatic forms, both plant and animal. Among the latter, the clams are the largest and most conspicuous, though probably not the most significant.

Aside from a few minute bivalves, hardly as large as a fingernail, in the whole series of the Okoboji lakes there are but two kinds or "species" of clams, (1) the "Fat Mucket" (*Lampsilis luteola*) and (2) the "Paper Shell" or "Floater" (*Anodonta grandis*). Though both approximate the same size, from two and a half to four inches in length, one to one and a half in width, and from one and a half to two and a half in depth, they may be distinguished quite easily. *Lampsilis* has a strong, heavy shell, with prominent scars and large interlocking "hinge teeth" within, and a fine luster to its pearly linings. Its exterior is bright yellow, well polished, often tinged with green and brown or shading into these colors, and frequently marked with radial streaks of green. In contrast, the *Anodonta* may be recognized by its exceedingly thin shell, so fragile that dead specimens are usually cracked or in fragments. It lacks the hinge teeth and has but faint scars. Its external color varies from a rather dull, pale straw color to a greenish brown and often in dead shells to a dark brown. The nacreous interior may show some irridescence when fresh, but it usually soon becomes dull and chalky.

Both species of clams are found in the three major lakes, and in all of them the *Lampsilis* is by far the most abundant. In the smaller adjacent lakes, clams are relatively scarce, and in most of them, the *Anodonta* alone is found. *Lampsilis* seems to prefer the shallower water, less than five feet in depth, and reaches its maximum abundance at about three feet. Firm, sandy bottom, or sand mixed with clay or compact gravel, either without vegetation or with rather sparse and strongly rooted plants, characterize the most populous situations. *Anodonta* is more abundant in deeper water, especially where the bottom is soft and somewhat muddy or even quite mucky.

In Spirit Lake, clams are abundant almost anywhere and everywhere where normal depth ranges from two and a half to six feet, except where the beach is quite rocky. Even here where boulders predominate at the shore line, the clams appear in the deeper water as soon as transition occurs to a softer bottom. The general uniformity of shore and bottom conditions in Spirit Lake has fostered a very wide distribution of clams and their occurrence in enormous numbers. Only in a few localities, such as sandbar ridges or along the western shore in deeper water, where soft bottom predominates, does the proportion of *Anodontas* rise even to a fourth the total number; elsewhere *Lampsilis* occurs in great excess.

The numbers and distribution of clams in West Lake Okoboji are in striking contrast to the conditions just cited for Spirit Lake. Opposed to the wide distribution and great abundance in Spirit Lake, in West Lake clams are to be found in particular locations only, and then in somewhat limited numbers. From some parts of the Lake they are quite generally absent. Presumably it is the highly varied character of the near shore bottom—and such variety is the most noteworthy feature of our West Lake—which determines the distribution and the frequency of the clams. A rocky, boulder beach is entirely destitute of clams, as is also a shifty or muddy bottom, or a shore where the slope to deep water is too abrupt. But where there are gentle to gradual slopes of fairly compacted sand or of sand mixed with moderately stiff clay or gravel, provided any attending vegetation is not overwhelming, clams are found at the head of the Lake, in Hayward's Bay, along the south shores of Emerson's Bay, and on the restricted beaches near Gull Point, the Manhattan, Egralharve, the outer side of the sand spit in

Miller's Bay, etc. The effect of bottom conditions in determining the location and distribution of the clams may be studied at Crescent Beach, for example. In the soft sand of the bathing beach just off the hotel, clams are infrequent; and along the east shore toward Bluff Point, where large rocks mixed with a stiff clay form the bottom, there are none. But between, where the sandy beach merges gradually into finer and then coarser firm gravel, the clams are lusty and in great numbers. Again, along the east shore of Smith's Bay, up near the drawbridge, where the vegetation is smotheringly compact, the clams are rare, but they increase to comparative abundance as one goes south toward Arnold's Park, as the number of aquatic plants decrease.

East Okoboji is almost completely destitute of clams except in a restricted locality along the beach at its head, just south of the State Fish Hatchery at Orleans, and for a very few scattered individuals found near its southern end. This very severe limitation in numbers must be surely due to the unsatisfactory bottom conditions throughout the Lake—a soft, shifting muck overlying boulders of generally large size. Curiously, however, those few clams which have found residence in such unfavorable bottom conditions are exceptionally large and vigorous. It is believed that their sturdiness is due to the great abundance of minute organisms carried in the waters of the East Lake—"plankton," the scientist names it—which gives to East Lake its consistent turbidity and greenish color, and incidentally furnishes exceedingly rich dietary for its clams.

It may be interesting to know that, on the whole, lakes do not furnish as favorable conditions for clams as moving streams. In the Little Sioux there are a half dozen other species in addition to those found in the Okoboji Lakes. And in these two species common to the two environments, the individuals taken in from the river are giants in comparison to their fellows from the lakes; and that in spite of the generally unfavorable bottom conditions in the Little Sioux, which resemble those in East Okoboji. To such inactive creatures as the clam, a constant current which shall bring in its food must be a factor in its success.

An interesting catastrophe occurred last summer to many clams in Spirit Lake. It will be remembered that lake levels had fallen consistently the last few years, so that the depth of the Lake had been reduced at least four feet below normal. The cyclonic storm of June 12th was particularly severe over Spirit Lake, and the force of the waves along the upper western shore, driven by the heavy and protracted blow from southeast, tore from their anchorage in the now shallow, soft bottom and tossed high upon the beach to die, by the conservative estimate of over a million clams. Just after the storm, from opposite Hottes Lake almost to Crandall's Lodge, the dead and decaying clams lay heaped in high continuous windrows. Counts made average nearly five hundred to the square yard. And curiously enough, these dead clams were almost without exception *Anodontas*. In the adjacent Lake, *Lampsilis* outnumbers *Anodontas* more than twenty to one, but they—active, big footed and thick shelled—could better "dig in" and withstand the fury of the storm; or even if washed out, they bore the pounding better. On account of their greater weight, they were not thrown so far from the water-line, and so succeeded in crawling back to safety. The *Anodontas*, though numerically much less common, furnished through their weakness the highly preponderating number of victims.

The writer's chief interest in the Okoboji clams has centered in the parasites from which they suffer. "As happy as a clam" is certainly a misnomer, as they suffer widely and curiously in this manner, certain parasites affecting to complete sterilization their reproductive organs, and almost no portion of the body being without parasitic injury. Four of the major divisions of animals are represented in the parasites of the Okoboji clams, and ten different kinds. Very interesting interrelations exist between the fish and the clams in this matter of parasitism, for in certain cases the clam plays the same intermediate place in the distribution of the fish parasite as poorly cooked infested pork and beef do in the case of human tapeworms. As the fish population of Spirit Lake differs from that of West Okoboji, as might be expected, there is a corresponding difference in the range of clam parasites in these two regions. And again, in the East Lake, lying geographically between Spirit Lake and West Okoboji and connected normally to both, and having an intermediate fish population, the features of its clams' parasitism are likewise intermediate.

WHY STUDY POND CRAYFISH?
Richard V. Bovbjerg

Aquatic Ecology was the course I taught for all those years at the Iowa Lakeside Lab (Ken Lang and Mike Lannoo are alumni). My favorite habitat, wetlands, were all around us. These ranged in size and permanence from small vernal ponds with water only in the spring, to huge sloughs with portions drying each year but completely drying only rarely. All these basins

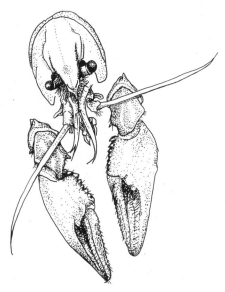

The pond crayfish Orconectes immunis. *About four inches long.*

have a single species of crayfish, *Orconectes immunis*. The other species in the region is the closely related *Orconectes virilis*, the stream crayfish, also found along rocky lake shores.

This stream crayfish just cannot survive the rigorous conditions found in wetlands. All sorts of experiments over many years were done in the lab to find the reasons, but the best experiments were done by nature. The Little Sioux River west of Milford flooded each spring over into a wetland, an oxbow pond. So in June the resident pond species mixed with the stream species that had washed in. By the middle of the summer only the pond species had survived. Why only the one species?

The wetland habitat is so harsh. In the winter the water is frozen and any water at the bottom is at zero oxygen and filled with noxious gases. By spring melt the bottom is dotted with the red carcasses of crayfish that did not make the winter. But the survivors greet the spring conditions of sunshine and exploding vegetation, which provides food (they eat anything) and oxygen in the water. The female lays eggs that glue to the underside of her abdomen; the eggs hatch, still holding to the mother. When the young drop from the mother they are on their own: the pond is in early summer splendor.

But even in the lush times the pond remains tough for crayfish. During the day they hide under plant cover. And while this dense green environment skyrockets the oxygen in the water to saturation or even supersaturation, during the night all that mass of life consumes oxygen down to zero by dawn. The pond crayfish can tolerate low oxygen for a longer time than the stream crayfish; specifically, long enough to last the night. During the night many crayfish leave the water for the shore; they prowl about feeding on living and dead vegetation, insects, snails, anything. In this dewy, moist air they can use their gills for gas exchange. They almost become land animals.

The ultimate rigor for wetland animals comes with the partial or complete drying later in the summer. And what of the pond crayfish? They leave the diminishing water for the shore where they dig burrows down to the water table. This behavior is lacking in the stream crayfish. Students and I monitored this behavior in a vernal pond; we called it Fairy Shrimp Pond— a classic, but now in row crops. A new ring of burrows was seen each day as the shoreline shrank. This was a rich pond, dense with vegetation, full of crayfish, frogs, salamanders, insects, and much more. Just before pond drying, the frogs emigrated. The salamanders had not left; they still had gills. The last stinking pool of water was their abattoir; they were essentially exterminated. The last salamanders fed birds and snakes; some snakes showed six or more salamander-sized lumps.

The crayfish escaped this fate. By the end, the shore was dotted with the mud chimneys of burrows. At the end of the summer we dug down to find the end of a single burrow. After digging a trench shoulder deep, we found a crayfish chamber at the now-deep water table. It had one active crayfish that resisted arrest. Such survivors had now a second year's growth (probably the last one), and they had mated and stored the sperm for the spring laying. The winter was spent at the water table below the frost penetration. Here they sat at a reduced metabolic state, in torpor.

We tried to study crayfish behavior and burrowing out in the sloughs. Field biology is exciting but also full of wet mud, biting insects, and frustration. We saw crayfish only occasionally. They were out at night; we saw none during the day.

Behavioral studies were moved into the laboratory. Our aquatic lab has large concrete tanks with air and lake-water taps. A "pond" was created. A tank of about six feet on a side was divided into halves: one half was a shallow water pond and the other half merged into a shore. The water was filled with pond vegetation and the shore with marsh plants of several kinds. The tank was enclosed by a black plastic blind with peep holes. A window admitted natural light; this was augmented by artificial light on a normal photoperiod. Groups of ten animals were marked and released into the pond and studied for a week. Individual positions were recorded at dawn and after dark (with checks during the day). Thousands of positions and behavior notes were taken during three summers. I can say the following things about behavior of pond crayfish under these conditions.

They are nocturnal animals. On land they prowl about, feeding, fighting, and mating. They return to the water at dawn, running into the water until completely submerged; no great blue heron beaks for them.

Surprisingly, an individual did not return to the same burrow it had made. At dawn, those on land might search for an unoccupied burrow. If all burrows were occupied, they could evict a resident or start a new one. The chimney was defended, but more aggressive animals could evict others. They did this by harassment until the occupant fled, not by assault.

Since the animals were marked, social dominance could be seen. Like other crayfish, larger animals dominated smaller; males dominated females. They would not tolerate a close neighbor; one animal per burrow was the rule.

In the lab I saw the actual construction of many burrows and chimneys. To start, a crude pit was formed at the surface. This provides some immediate cover. Then the burrow was made by jabbing into the mud with the claws and pulling the mud back (like a back-hoe machine). Pellets were formed of the soft mud and pushed into a ring around the hole; a chimney was started. The tunnel was dug down to the water, only a few inches in the lab tank. The burrow was then enlarged into a chamber with water. Now any further excavation was done by pushing mud to the surface (more like a bulldozer). The chimney was an imperfectly round mound with a hole in the center that fit the crayfish builder.

An existing burrow was always entered head first in exploration. The animal then backed up, turned around, and descended tail first in defensive posture. Finding lateral burrows into the bank was a surprise; of course, we never could have seen that in the field.

Juveniles in June never left the water. They hid in vegetation. By July they started to emerge on land in the night. Now they were able to dig tiny burrows just below the surface, clearly an unlearned and complex behavior, an instinct.

The stimulus to emigrate was the drying of the lab pond. At the end of each week, I siphoned off the water. One hundred percent of the animals headed for shore, even at mid-day. They instantly burrowed. However, out in a natural pond burrowing starts before complete drying. How do they know to do this? Nice question: photoperiod, temperature, noxious water? More experiments?

This procedure of extensive field studies combined with laboratory observations and experiments has scientific power for probing ecological questions.

WHY STUDY POND SNAILS?
Richard V. Bovbjerg

Wetlands are just the right place to collect a few snails for your home aquarium. Any animal from there would be at home in an aquarium's stagnant water. The giant pond snail (large as your thumb) is indeed named *Lymnea stagnalis*. What are the other wetland mollusks? How do they survive there? Are there distribution patterns between and within wetlands? How do we know?

Clams and snails are the two types of mollusks in the Okoboji region. Clams have two hinged shells; snails have one coiled shell. Locomotion is slow and done with a muscular foot. In clams the foot digs into the bottom and pulls the animal. The snail's foot is flat, and it allows the animal to creep on plants and other surfaces such as fallen branches or even the underside of the water surface.

There are few clam species in wetlands; there are the tiny fingernail clams (the size of a small child's finger). However, these may be abundant, hundreds in a square foot of pond bottom. There are more species of snails in a pond, maybe a dozen. Some snails have gills and get oxygen from the water, as fish do. Most pond snails of the region are pulmonates. These breathe air and retain it in a simple sort of lung, a vascular membrane lining a chamber inside the shell. Like tadpoles and salamander larvae, they must rise to the surface and open this chamber to exchange air.

Many of the wetland mollusks are not found in lakes and streams. But even within wetlands, different basins will have different species. Some species might be rare in one pond and abundant in another. In some cases I could identify the pond a particular snail came from. Ponds are not identical nor are the mollusks in them. In the pools of Silver Lake Fen there is one species of snail, and it is not found in other ponds nearby. The giant pond snail is most abundant in one part of Jemmerson Slough that is crossed by

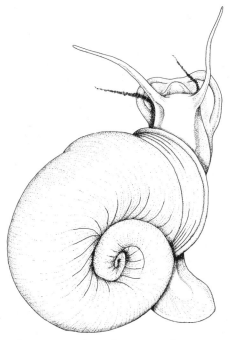

The pond snail Helisoma trivolvis. *Diameter about one inch.*

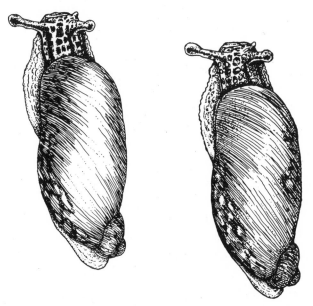

Amber land snail, Oxyloma retusa, *found along the damp borders of Okoboji's wetlands and rivers. About one-half inch long.*

the railroad tracks. A planorbid snail, flat and disclike, is most common in the wetlands west of Spirit Lake. It cannot be said of wetlands that when you have seen one you have seen them all.

It is not surprising that ecological rules of islands apply to ponds, which may be viewed as islands in the sea of grass (and crops) in this region. The larger the wetland, with more diverse environments, the greater the species diversity. Small vernal ponds support a couple of snail species; complex, large sloughs support many times more.

The question still remains, Why is one species found in a particular place? Why the patterns? Any mollusk population is where it is for the following obvious reasons: (1) it had the capacity to get there; (2) the needs of the animals were met there; (3) the extremes of the environment were tolerated there. Always, chance plays a role in pattern.

Moving from pond to pond is impossible for a creeping aquatic snail but would not be surprising for a beaver, muskrat, raccoon, or marsh bird. So these are the very agents most ecologists suspect are carrying mollusks, especially tiny ones or eggs. They could be carried some distance in muddy fur or feathers. Lots of exciting studies are yet to be made of this dispersal mechanism.

By whatever means a snail arrives at a new pond, the special needs of that animal must be met or it will not survive. A primary need for all shelled mollusks is calcium carbonate, abundant in the Okoboji region. In acid bogs there are no mollusks at all; there is no calcium in acidic waters. Food is seldom a problem; it is everywhere in wetlands. Fingernail clams extend their siphons up from the ooze and filter the microscopic life and particles from the water. Snails eat plants, living or dead; they scrape food with a rasplike tongue. The need for cover from predators is met, for wetlands are lush with plant cover. Pulmonate snails get oxygen directly from the air, so the oxygen needs are met even when the aeration of the pond is low.

And what are the extremes they must tolerate in wetlands? The harshest times are winter freezing and summer drying. Like almost all animals of wetlands, mollusks respond by burrowing into the bottom. Here they become inactive and secrete protective layers of mucous to retain moisture. My students and I have brought chips of cracked mud from dried ponds back to our laboratory. We flooded this dry mud with clean water and overnight we found the water teeming with life, mostly small forms, but including snails. Survival of pond animals is dependent on this capacity to stay in a low metabolic state; they seem dead but are capable of revival.

Giant conical pond snail, Lymnea stagnalis, *depicted crawling upside down on the water's surface. Two inches long.*

This recitation of traits points to all aspects of wetland mollusks: anatomical, physiological, and behavioral. Mollusks are classical examples of adaptation to survival in this habitat.

But we can see a further pattern; even within one wetland the dispersal of snails is not random. There is pattern within ponds, repeated in pond after pond. I noticed that the giant pond snail at Jemmerson Slough was not at the shallow margins, not in the open center; they were in knee-deep water in a zone of dense vegetation. At isolated spots they were crowded around pieces of animal carrion, such as dead insects or crayfish. How do they find these? Slow as they are, mollusks can move; they select their habitat. They must respond to environmental cues positively or negatively, approach or avoid. The sensory region is the head and foot, but especially the pair of tentacles extending forward. This is a mindless response, a reflex; the brain for the mouth region is only about a couple of dozen nerve cells.

In the laboratory we tried to sort out this sensory-response reaction. Experiments were done on perception of foods and the responses. Snails detected both plant and animal cues and started feeding responses. But only animal food was detected at any distance. They followed a chemical gradient to the source.

Eventually, I got around to doing experiments with multiple and simultaneous factors. This was a very simplified pond simulation in a pan. Three factors were varied, one at a time, then in pairs, and finally all three. This little environment had a warm corner and a cool corner, areas with and without vegetation, and four small spots of crayfish carrion. The snail positions were mapped. In the final three-factor experiment, the dots on the map became a solid patch at the cool corner with carrion. It was repeated many times, with the same pattern. Snails responded synergistically to multiple cues and vectored toward the optimum.

As it is with pond snails, pattern is the rule in nature, with the usual caveat not to rule out chance events. Pattern in nature is dictated by ecological cause and effect. So the next time you see a willow tree sunning on a river bank or a monarch butterfly drifting about in a roadside ditch, ask them why they are there at that exact spot. They cannot answer; only you are blessed with inquiry.

AMPHIBIANS OF OKOBOJI

While many invertebrate groups inhabit Okoboji's wetlands, as a whole no vertebrate group is as dependent on natural wetlands as are amphibians. Not only do amphibians need wetlands, but wetlands—to function properly—need amphibians. Wetlands are not just small lakes. Wetlands have characteristics that distinguish them from lakes, characteristics that make them unique. Much of this uniqueness relates directly or indirectly to their suitability for amphibians.

The origin of the modern amphibians is obscure. Ancient amphibians arose about 300 million years ago and, despite their delicate appearance, have persisted. They survived the dinosaurs, and whatever killed the dinosaurs. They may be losing the battle to survive the environmental impacts of humans.

Today, modern amphibians are grouped into three orders: frogs (Anura); salamanders (Caudata); and caecilians (Apoda). Okoboji has no caecilians, a small group of legless, burrowing species located in Central and South America, Africa, India, and Southeast Asia. Okoboji's six species of frogs represent three of the twenty-two anuran families. American and Great Plains toads are in the family Bufonidae. Chorus frogs and gray treefrogs are treefrogs in the family Hylidae. Leopard frogs and bullfrogs are so-called true frogs in the family Ranidae. Okoboji's only extant salamander, the tiger salamander, is in the mole salamander family Ambystomatidae.

OKOBOJI'S HISTORICAL AMPHIBIAN FAUNA

In 1920, the Michigan herpetologist Frank Blanchard visited Lakeside Lab for five weeks and conducted Okoboji's first amphibian survey. He found the following amphibians: mudpuppies (*Necturus maculosus*, based on one or two specimens); tiger salamanders (*Ambystoma tigrinum*); chorus frogs (*Pseudacris triseriata*); gray treefrogs (*Hyla versicolor*); cricket frogs (*Acris crepitans*); leopard frogs (*Rana pipiens*); and American toads (*Bufo americanus*). Blanchard found tiger salamanders to be abundant: "at least one specimen was found in nearly every pond seined." He also found cricket frogs to be "common but not abundant." Leopard frogs were the "widespread and abundant amphibian of the region. Specimens were taken in all the townships visited. They occur near all the lakes, sloughs, and streams and [are] common in tall and sometimes in short grass, on the prairies and uplands far from water."

From Blanchard's data we consider that in 1920, leopard frogs and tiger salamanders were abundant (i.e., found in most wetlands sampled), the American toad, chorus frog, and cricket frog were common (found in several sites), and the gray treefrog and mudpuppy were uncommon (found in a few sites).

During the summers of 1991, 1992, and 1993, we repeated Blanchard's survey and, like Blanchard, found seven species of amphibians in Dickinson County. Interestingly, we did not find exactly the same species. We found the eastern tiger salamander, the American toad, the Great Plains toad (*Bufo cognatus*), the western chorus frog, the gray treefrog, the northern leopard frog, and the bullfrog (*Rana catesbeiana*). The following essays detail the characteristics of each of our amphibians, and the 1990s amphibian assemblage is then compared to the 1920s assemblage.

TIGER SALAMANDERS

Like frogs but unlike mudpuppies, tiger salamanders have what biologists call a complex life history, meaning they spend part of their life as an aquatic animal and part of their life as a terrestrial animal. Each spring around the end of March or early April, soon after the ice has melted off the wetlands, terrestrial adult tiger salamanders migrate from their overwintering sites in burrows and under rocks and logs to their breeding ponds. Males—easily identified by their swollen cloacal glands—enter the water first and wait for females. Females—bulging with eggs—are courted after they enter the pond. Males interest females through a tail-waving and body-undulating display. Females show acceptance by following the males. The male excretes a small bundle of sperm known as a spermatophore. The female, following, passes over and picks up the spermatophore. Fertilization is internal.

LARVAE

At hatching, tiger salamander larvae are between a quarter and a half inch long. Larvae are a mottled green to gray color, with a lighter belly. They have a tail with a fin and external gills projecting out from behind their head. As they grow, larvae develop front limbs first, which may help stabilize them as they swim and feed—two behaviors that begin within a few days after hatching.

Larval tiger salamanders are typically the top aquatic carnivores in Okoboji wetlands. They swallow their food whole and will eat any animal they can fit into their mouth, including other tiger salamander larvae.

Salamander larvae are carnivores. As they grow, the range of prey sizes they can eat increases, and as a consequence the range of prey types they can eat expands. The youngest (i.e., smallest) larvae eat microscopic zooplankton such as the water flea (*Daphnia* species). As salamander larvae grow, they continue to eat *Daphnia*, but add aquatic insect larvae, aquatic worms, snails, crayfish, frog tadpoles, and each other to their diet.

Salamander larvae live about ten weeks before metamorphosing into adults. In contrast, frog tadpoles are herbivores and can metamorphose much more quickly, about seven weeks or less. Tadpoles feed by scraping algae off submerged rocks and plants. Part of the reason tadpoles metamorphose more quickly is that, unlike salamander larvae, they do not exert much energy searching for, or chasing after, prey. This is basically the same reason that buffalo grow faster (and bigger) than wolves. The tradeoff is that nutrients are more difficult to extract from consumed plant tissues than from animal tissues. To access these nutrients, tadpoles need more gut—

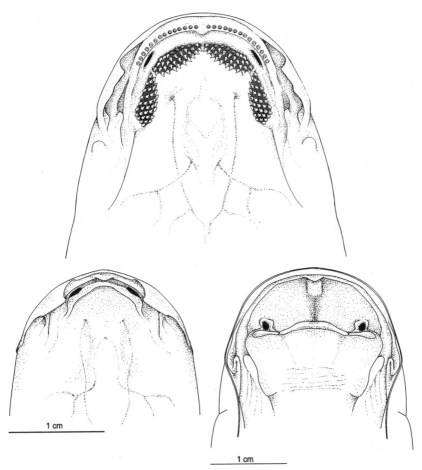

Schematic views of dissections of the roofs of the mouths of tiger salamanders.
Top: cannibal morph larva; bottom, left: typical morph larva; bottom, right:
adult.

more intestinal surface area—and so they increase intestinal length and coil
it like a watchmaker's spring. Their fast growth attests to the success of their
modified intestine. But there is a cost. Unlike salamander larvae, which essentially have an adult gut, the tadpole gut must be entirely rebuilt during
metamorphosis. This rebuilding includes the construction of a stomach. Albert Kuntz, working at Lakeside Lab in 1924, provided an early description
of the metamorphic changes in the gut of the leopard frog and compared
them to changes in tiger salamanders. Adult frogs are carnivores and have a
gut similar to both adult and larval salamanders.

SPECIAL SENSES

Tiger salamander larvae are nocturnal sit-and-wait predators. When feeding they tend to hang motionless in the water column and wait for prey to come toward them. How do tiger salamander larvae sense prey at night? My own experimental studies have shown that tiger salamander larvae do not need vision to feed normally at night. In fact, their eyes are tiny and their prey so small and transparent that they probably cannot use their eyesight to feed nocturnally. Smell is also not an important sense for feeding tiger salamander larvae. Even their slowest prey, *Daphnia*, can swim much faster than scent molecules can diffuse through water. By the time a salamander smells a *Daphnia*, it has probably swum out of striking distance.

Like fish, tiger salamander larvae have a lateral line system. In fact, the lateral line of salamanders is much more sophisticated than that of most of our fishes. The lateral line of most Okoboji fish consists only of mechanoreceptors. Tiger salamander larvae not only have mechanoreceptors but also electroreceptors. Mechanoreceptors allow motionless salamanders to detect the water currents generated by prey and predators. Electroreceptors detect electrical currents generated by an animal's muscle contractions—for example, when their heart beats, or when they are swimming. Tadpoles resemble our fishes in that they possess lateral line mechanoreceptors, but not electroreceptors.

Two native fishes, the paddlefish, now extirpated in Okoboji but present in other regions, and the various catfishes, also have electroreceptors. Electroreceptors in paddlefish are especially dense along their paddle. Unfortunately, paddlefish electroreceptors did not enable them to sense the danger from the nets of Okoboji's turn-of-the-century commercial fishermen.

CANNIBAL MORPHS

We have an unusual animal in Okoboji. All our other animals are found elsewhere except our tiger salamanders, which have a unique combination of features: adults with yellow spots; adults that lay their eggs in clusters; larvae that have fourteen to twenty gill supports; and larvae that exhibit a cannibalistic morphology.

During droughts, wetlands dry sooner than normal. Low water levels concentrate tiger salamanders and crowd them. Under these conditions, a small percentage of larvae develop large heads and large teeth on the roof of their mouths and on their upper and lower jaws. Instead of concentrating

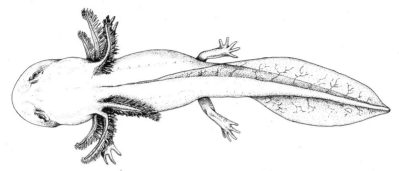

Cannibal morph eastern tiger salamander larva, an animal known at this time to occur naturally only in Okoboji. Note its relatively large head. Up to about five inches long.

their feeding on *Daphnia* and eating large numbers of these small prey, cannibal morph larvae shift their diet and eat small numbers of large prey. The diets of these larvae include frog tadpoles, small crayfish, dragonfly larvae, and other salamander larvae. Why?

It is tempting to conclude that tiger salamander larvae become cannibalistic during drought conditions to insure that at least some salamanders in each wetland will survive. For amphibians, larvae typically do not metamorphose until they reach a certain critical, threshold size. Cannibals have an opportunity to reach this size sooner because they eat not only aquatic invertebrates but each other. There are simply more kinds of prey to eat, so prey are encountered sooner. It is also possible that cannibals get an additional dose of growth hormone (thyroxine) for each salamander eaten, which may further accelerate their growth. The problem with this scenario is that we know that natural selection acts not on groups or populations but on individuals. Arguments based on "for the good of the population" logic do not usually work in a Darwinian world. The advantage of cannibalism to the salamander cannibal is not shared by their salamander prey. It is not in an individual tiger salamander's best interest to be eaten, no matter who is doing the eating.

Cannibal morphs are not often found during normal water years (but can be induced in the laboratory under crowded conditions). In nature, therefore, some Okoboji tiger salamander larvae have the genetic potential to become cannibal morphs but do not express these genes unless they experience environmental stress. When cannibal morphs do occur, typical morph larvae persist. Therefore, individuals within populations of tiger salamander larvae can exhibit different shapes. Biologists call this polymorphism. Cannibal morphs are an example of an environmentally induced, genetically determined polymorphism.

WETLAND ANOXIA

If wetlands go anoxic and tiger salamander larvae have gills like fish, how can they breathe? Tiger salamander larvae experience the conditions of summerkill—nighttime anoxia. But they do not suffocate. In addition to having external gills, larval tiger salamanders, like all of Okoboji's amphibian larvae, have lungs. When oxygen levels drop, larvae either swim or crawl through the submergent plants to the water surface, part the duckweeds with their snout, and gulp air, taking oxygen into their lungs (an inspirational behavior).

In addition to having lungs, amphibian larvae can breathe through their skin. They can sometimes be seen lying near the surface of the water, underneath the duckweed. Here they absorb the atmospheric oxygen that diffuses into the extreme upper region of the water column.

BUOYANCY

Lungs not only enable tiger salamander larvae to breathe under conditions of low oxygen but also allow them to suspend themselves in the water column. Inflated lungs serve as hydrostatic organs and allow salamanders to become buoyant with little effort or energy expended.

FEEDING

Buoyancy is important to Okoboji tiger salamander larvae because they feed at night up in the water column. By far, the predominant prey of tiger salamander larvae are *Daphnia*, the tiny (one-twentieth inch long) water fleas. Newly hatched salamanders feed on *Daphnia*, and while they add other prey types to their diet as they grow, they never lose their habit of eating zooplankton. In one of my studies, eighty-two salamander larvae had an average of two hundred and fifty *Daphnia* in their stomachs. One larva had recently eaten over ten thousand individual *Daphnia*.

To eat, tiger salamander larvae open their mouths and suck in, taking in the prey and water around it. No matter how large or small the prey, tiger salamanders swallow it whole; they do not bite off chunks, they do not

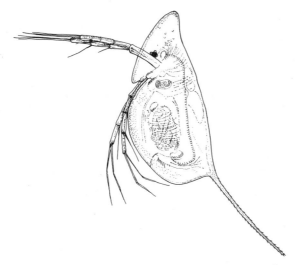

The zooplankter Daphnia longiremus, *a native West Lake Okoboji species now believed to be extirpated. About one-twentieth of an inch long.*

chew to break up food. They hold the prey in their mouths and expel the water out from openings behind their gills. To be able to take prey ranging in size from microscopic *Daphnia* to each other, they have specialized mouths. Along the sides of their mouths they have flaps of skin, called labial lobes, that are capable of stretching. To take a tiny *Daphnia*, they open their mouths slightly and suck. The flaps of skin reduce their mouth opening and make it work like the end of a straw—by restricting the size of the opening the suction becomes more forceful. To take larger prey, tiger salamanders open their mouths wider. To take the largest prey, they bite and hold on, and their mouth flaps expand to accommodate the bulk.

LOCOMOTION

The versatility of tiger salamander larvae extends beyond the way that they feed and breathe to the way they move. Larvae have tail fins, which they use to swim, and legs, which they use to walk. Swimming is used for fast locomotion. In pure swimming, the body undulates like a fish, and the limbs and gills are compressed against the body. To stop swimming, the limbs are extended and used as brakes. Limbs are used for walking or in assisting larvae as they work their way through the tangles of submergent vegetation.

Adult eastern tiger salamander, Ambystoma tigrinum tigrinum. *About seven inches long.*

METAMORPHOSIS AND ADULT ECOLOGY

In Okoboji, tiger salamanders remain larvae for about ten weeks—seventy days—and metamorphose in early to middle July. This period may be shorter in warmer years or during drought years.

Metamorphosis involves a major restructuring of the larval body. The external gills, mouth flaps, and tail fins are resorbed. Limbs become stronger to support the body on land. The skin becomes thicker and forms yellow or bronze-colored spots. The openings in the neck region underneath the gills are sealed, and the gill supports are resorbed. At this time cannibal morphs can no longer be distinguished from the adults of typical larvae.

Tiger salamanders within a wetland tend to metamorphose at about the same time. One rainy night in July the fields and the roads come alive with juvenile tiger salamanders. In the past, before the weir was built connecting Garlock Slough to West Lake Okoboji, metamorphosing salamanders would make a hazard of the Highway 86 curve near Shuck's Corner. Car-crushed salamanders made the pavement slick with guts. Old timers will remember this.

Dick and Ann Bovbjerg (1964) write about leopard frogs exhibiting a similar mass metamorphosis: "During the evening of 8 July 1961, a massive swarm of juvenile frogs left the waters of Garlock Slough in Dickinson

County, Iowa, and moved into the surrounding hills of grass and cropland so that the tall grass was alive with small frogs. They crossed an adjacent highway [Iowa Highway 86] in such numbers that automobiles crushed a carpet of frog viscera. This emigration continued at a decreasing rate for the next two weeks." Today, no amphibians regularly breed in Garlock except for introduced bullfrogs.

After metamorphosis, juvenile tiger salamanders disperse to the woodlands and the prairies. There they are secretive, burrowing into the ground or under rocks and logs. Salamanders make their living eating earthworms and insects, including agricultural and horticultural pests. Although adult salamanders are not obvious to us, salamander biomass can equal or exceed the combined biomass of all birds and mammals. Tiger salamanders are capable of extensive dispersal; they have been found as much as ten miles from their home wetland. Adults can live a long time, more than a decade in captivity.

Tiger salamanders hibernate during the winter. Each spring they emerge and migrate to wetlands to breed. Mortality must be high for salamanders. Each female can produce about one thousand eggs per year. Therefore, over the course of ten years, females potentially lay about ten thousand eggs. But we are not in danger of being overrun by salamanders. If the number of salamanders remains constant, each female on average will produce just two breeding adults in her lifetime. Therefore, the chance that a given egg will produce a breeding adult is low, perhaps one in five thousand.

Adult salamanders provide food for the predatory birds and mammals of Okoboji. Hawks, herons, pelicans, raccoons, opossums, badgers, coyote, and fox all feed on salamanders at some point during the year. Garter snakes congregate around drying wetlands to feed on the easy pickings of trapped amphibian larvae.

AMERICAN TOADS

Of the two Okoboji toads, the American toad is by far the most numerous. American toads are amphibians of the east; Okoboji is near the western edge of their range. The call of the American toad is a long "trill." American toad tadpoles are tiny, black, and may form schools. They metamorphose

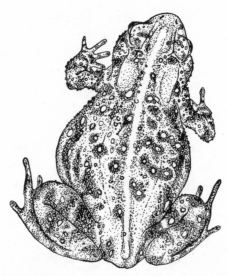

Adult American toad, Bufo americanus, *which has small, dark pigment spots usually containing one to three warts. About three inches long.*

into miniature replicas of the adults. Whereas most other Okoboji amphibians do much of their growing as tadpoles—reaching sizes ranging from perhaps thirty to sixty percent of their adult size—newly metamorphosed toads are only a small fraction of their adult size. Portions of the following description are drawn from Mary Dickerson (1908).

The color of American toad adults is variable, usually yellow, tan, or brown; a light midline stripe may be present. The skin is noticeably warty. Warts may be bright red or yellow. The undersurface is light. The male's throat is black. Females may be brighter and more varied in color than males. Males range from two and a half to three inches in length; females are larger.

American toads are in the ponds in May and June, appearing earlier during warmer springs. The males sing and are the first to arrive. Females arrive and choose a mate. Fertilization is external—the males release their sperm over the eggs as they are being shed. The eggs are laid in long strings that are found wound around submerged plants. Eggs are small (less than one-eighth inch in diameter), black above, white below, and arranged in a single row in a transparent jellylike mass. Their number is enormous—between four and twelve thousand from a single female.

In as little as four days from the time of laying, the tiny black tadpoles have hatched and can be found clinging to the jelly mass or to nearby plants. Tadpoles adhere by a sticky substance produced by two suckers near their mouth. Two days later, the head, body, and tail are more clearly differentiated from one another, and external gills are present. Tadpoles are more active and begin to swim in short circles. A few days after this the tadpoles are "veritable pollywogs": larger, coal black, actively swimming, and beginning to feed. Mouths are provided with keratinized (the protein of fingernails) denticles for scraping attached algae from rocks and larger plants.

In about three or four weeks the full-grown tadpole is less than an inch long, with the tail more than half of this length. The color is nearly black, but when we look closely—especially at the iris of the eye—we can see gold stippling. Its belly may be colored bronze.

Before metamorphosis to the adult toad, the tadpole develops hind legs. The forelimbs are also developing but are tucked underneath the operculum, the flap of skin that covers the gills. At metamorphosis the front limbs break through this skin, and the animal looks like a tiny toad with a long tail. Furthermore, at metamorphosis the mouth quickly undergoes a massive change. The denticles get resorbed, the jaw and gut get reworked, and the mouth becomes much larger. The eyes become larger and protrude. The tail becomes shorter until it is finally gone. Toad tadpoles metamorphose together, and for a few days in mid-to-late June the wet margins of Okoboji's wetlands are alive with tiny baby toads.

The skin of toads is slightly poisonous, and when a toad is grabbed it "pours out this fluid in sufficient quantity to cause it to appear in milky drops on the gland-like swellings. This fluid has a disagreeable effect on the mucous membranes of the mouth, and so protects the toad from many enemies." Indeed, when you see a dog playing with a toad, it is nearly always a puppy. Even playful older dogs wisely avoid American toads.

Small lights mark the trails on the Lakeside Laboratory grounds. These lights attract insects, especially caddis flies, emerging from the lake. Under many of these lights you will find a single large adult toad. The lights draw the insects, the insects draw the toads. Why only a single toad? One answer may be that they are territorial and possessive of feeding sites.

Field sketch of the Great Plains toad, Bufo cognatus, *which characteristically has many warts on its large, green pigment spots. About two inches long.*

GREAT PLAINS TOADS

We freely admit that we do not know much about Okoboji's Great Plains toads; in fact, nobody does. We at Lakeside Lab are the only people we know of who have ever seen them alive in Okoboji, and we have seen only five of them. The first Great Plains toad I encountered was in 1991 with my class on the prairie path from the parking lot to the kettlehole at the Freda Hafner Preserve. In a matter of seconds after detecting us, this large toad had burrowed into the hard pan of the path and had covered itself completely. I scooped it out. The animal had large green spots on its back with cream-colored—almost yellow—borders. In color and ability to bust prairie sod, this is the John Deere plow of toads. A second Great Plains toad was found in 1991 on Cayler Prairie. Two more were found in 1992 on the Lakeside Lab grounds. One of these was found at night inside a well-lit lab building—drawn to the light (or, more likely, the insects). At first, we found Great Plains toads; then they found us. We do not yet know where they breed, although one student claims to have found a tadpole in the Little Sioux River.

According to Albert H. and Anna A. Wright (1933), Great Plains toads have a light tan to brown base ground color. Usually, four pairs of bright green spots outlined in white are present down the back. The underparts, including the throat, are light. Great Plains toads are said to breed from May to July in the northern states. Little is known about the eggs and tadpoles.

CHORUS FROGS

Chorus frogs are treefrogs and at about an inch long are Okoboji's smallest amphibians. They are brown or tan and usually have three darker brown stripes down their back. They have a white upper lip. Their song sounds like that made when you run your fingernail along the teeth of a plastic comb. There are several types of chorus frogs found over much of North America. Chorus frogs seem to tolerate agricultural practices well and may be as numerous in Okoboji now as in 1920. Why did Okoboji's cricket frogs decline and not its chorus frogs? Did chorus frogs benefit from the cricket frog's decline? Portions of the following description are drawn from Mary Dickerson (1908).

Chorus frogs can be found almost anywhere during the summer from uplands to wetland edges, but they are so small and camouflaged that they are seldom seen. They feed upon various insects including agricultural pests. Choruses are most robust in April and May but can be heard at anytime throughout the summer following heavy rains that fill wetlands. At a distance the chorus frog chorus is soft "and is said to have a soothing sound that swells and recedes like the waves of the seashore." Up close, you are amazed that such a loud sound can come from such a small animal. Only males call, and, when calling, their inflated throat pouch is large.

The eggs are laid in wetlands, usually in April or early May. They occur in small bunches of from five to twenty eggs and are attached to the leaves and stems of submerged plants. At the time of hatching, the tadpoles are nearly black in color; later, they turn dark green. Older chorus frog tadpoles

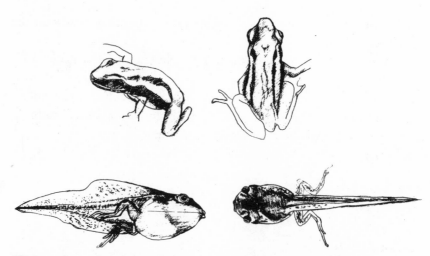

Western chorus frog, Pseudacris triseriata triseriata, *Okoboji's smallest amphibian. Total lengths: adults, one-half to three-quarters of an inch; tadpoles, about one and one-quarter inches.*

can be distinguished from those of the leopard frog, which hatch at about the same time, by their smaller size (maximum length about an inch and a half), squarer head, and sharply pointed tail. Chorus frogs are not good swimmers, and animals in the later stages of metamorphosis will drown unless they are supported by vegetation or move onto land.

GRAY TREEFROGS

Okoboji's most endangered amphibian (if, as we suspect, the cricket frog is locally extinct) is the gray treefrog. The gray treefrog is an animal of the eastern deciduous forest; Okoboji lies near the western limit of its range. In fact, at Okoboji's latitude we may have the westernmost populations. Gray treefrogs are common along the west fork of the Des Moines River, in Emmet County.

Since the deforestation following settlement, the gray treefrog may have never been numerous. Blanchard found two populations, one along the Little Sioux River and a second on the Kettleson Hogsback grounds. The Little Sioux population is now gone. In 1991, a single male gray treefrog was

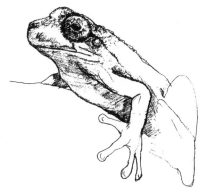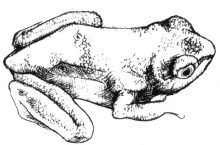

Adult gray treefrog, Hyla versicolor, *Okoboji's rarest extant amphibian. About two inches long.*

heard calling at Kettleson Hogsback. No males were heard there in 1992, 1993, or 1994. We have heard that a third population exists on the east side of East Okoboji, on the Camp Foster grounds. Our attempts to confirm this have been unsuccessful. In fact, the only wetland on the Camp Foster grounds is packed with bullfrog tadpoles. In 1994, we collected, photographed, and released a single calling male gray treefrog from a flooded roadside ditch in Gull Point State Park. In 1995, three males called from the state park. We have yet to find tadpoles.

The gray treefrog can be gray colored, but it also may be green, brownish, or some combination of these colors. The best way to distinguish it is by its large adhesive toe pads and the yellow-orange color on the inside of its hindlegs.

The gray treefrog forms two species that are together spread across the eastern United States. These two species are distinguishable only by call, or, if you want to get fancy, by chromosome count. The Okoboji species, *Hyla versicolor,* has twice the number of chromosomes as its sister species, *Hyla chrysoscelis.* The versicolor call is the lower pitched and slower of the two, although it is not always easy to distinguish them, especially when the weather is cool. Cold temperatures slow the metabolism of amphibians, reducing their call speed and lowering its frequency. Once more, Mary Dickerson (1908) provides some of the details in the following description.

At a distance the gray treefrog call sounds something like the bleating of a sheep. The pitch is uniform but may vary with the individual and continues for two or three seconds at a time, then ends abruptly. It is usually given several times in close succession.

Eggs are attached singly or in small groups to plants at the surface of the water. They are not easily found unless these plants are separated and examined minutely. The eggs are light colored, gray above, white below. They hatch on the second or third day. Newly hatched tadpoles are light yellow and about one-fourth inch long. Larval development proceeds rapidly; in three weeks the tadpoles have hind limbs. Tadpoles are a metallic gold, mottled with black, with red eyes and a red or orange tail. In about seven weeks metamorphosis is complete and the animals leave the water.

Hyla versicolor adults have a strong site fidelity and will often remain in one place for weeks or more at a time. In fact, it is said that these treefrogs can stay in a single tree for months.

CRICKET FROGS

Blanchard's cricket frog is found in the Midwest and eastern Great Plains. Two other subspecies exist, the coastal and the northern. The coastal cricket frog is found in a restricted range along the Gulf of Mexico in Texas and Louisiana. The northern cricket frog is misnamed; it occurs in the southern and eastern United States.

We could not find cricket frogs in our survey of thirty-four Dickinson County wetlands, despite sampling areas where Blanchard reported finding them. Are they gone? If so, why did they disappear? What killed them: lack of habitat, pesticides, reductions in prey, predation from bullfrogs? Will they come back? Aside from preserved museum specimens at the University of Michigan's Museum of Zoology, we have never seen Okoboji's cricket frogs and so must rely on Mary Dickerson's (1908) description more than usual.

The color of the cricket frog is variable, usually some shade of brown. The diagnostic feature is a triangular dark mark between the eyes and pointing backwards. On its back, spots may be green or red-brown and are often outlined with light. The eye is bright orange. The throat of the male in spring is light yellow. Underparts are light colored.

The size of the cricket frog is about an inch to an inch and a half in length—between the size of the chorus frog and the gray treefrog. The skin is rough and warty. Hind legs are long, the foot is fully webbed, toe discs are tiny.

The cricket frog is (was) the most conspicuously active of our small tree frogs. Despite its family name, it cannot climb shrubs and trees to get out of danger. The cricket frog remains on the ground throughout the year, preferably along the muddy margins of wetlands and rivers. When frightened, it jumps high and far, repeating these leaps in remarkably rapid succession. If it is disturbed near the water, it gives one or more of these remarkable leaps, swims vigorously a few strokes—using to good purpose the large webs between its toes—and buries itself in the bottom muck.

Cricket frogs are named for their call, which (it probably goes without saying) sounds like the chirping of a cricket. Only the male sings, its yellow throat inflated. Cricket frogs are easily found while they are singing; they do not hide under plants but call in full view on some water plant or floating twig. Cricket frogs breed late. Their chorus is loudest in early summer, and it is then that the eggs are laid, attached to grass blades or leaves in the water. The development of this frog is less rapid than that of most other frogs, and tadpoles may be found in the water as late as August. Metamorphosis occurs in early autumn.

LEOPARD FROGS

Leopard frogs remain the most numerous Okoboji amphibian. Several inches long, either green or brown colored (they do not change colors with background) and usually with dark spots across their back, leopard frogs are the frogs most people envision when asked to imagine a frog. Walk up to an Okoboji wetland and you will be preceded by the sound of water splashes, evidence of leopard frogs. When frightened, they jump into the water and hide by submerging themselves.

The same leopard frogs found in Okoboji are found across much of the northern half of the United States. They inhabit most of Iowa, except for parts of the south central portion of the state. While they remain widespread, there is evidence that their numbers are declining. We now know that for every leopard frog alive in Okoboji today, there were a thousand alive at the turn of the century. Imagine University of Iowa's Kinnick Stadium full, then with seventy-five people in it—this is the magnitude of our loss.

Portions of the following description are drawn from Mary Dickerson (1908).

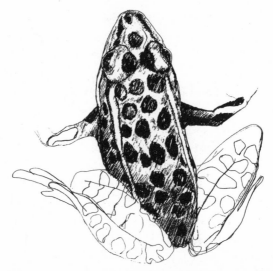

Northern leopard frog, Rana pipiens pipiens, *Okoboji's most numerous amphibian. Three to four inches long.*

The leopard frog is better known than others, not only because of its wider distribution and greater numbers, but because it occurs in uplands, traveling considerable distances from its home wetland. It is the frog met when we walk through prairies and fields. These frogs leap out from underfoot. They make long, low leaps, changing direction with every jump.

Leopard frogs are among the first of our frogs to come from their hibernation in the spring. Wetlands resound with their voices during the second half of March and early April. Eggs are usually laid in masses and may be attached to plants or left free in the water. Tadpoles hatch after about a week and a half to two weeks. The tadpoles are as black as toad tadpoles at the same age. During the next few weeks the body becomes brown or green and increases greatly in size to a maximum of more than three inches. Their life as tadpoles, which lasts about seven weeks in Okoboji, seems to consist of only four needs: to swim rapidly, to eat almost constantly, to rest a little sometimes, and to grow. The tail with its broad transparent fin is nearly twice as long as the rest of the creature.

In July, borders of Okoboji's wetlands are swarming with small frogs. Their home is in the shallow water among the submergent plants or along the wetland margin. Disturb them on land and they jump in enormously long, low leaps, into the water. It is nearly impossible to catch them by hand.

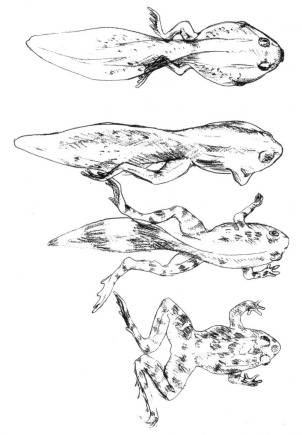

A series of leopard frogs representing stages through metamorphosis from late tadpole (top) to juvenile (below).

BULLFROGS

The bullfrog is an aggressive predator, a nasty piece of business, that will eat nearly every vertebrate smaller than itself. Sure, it provides larger frog's legs than leopard frogs, but at what cost? Our present bullfrogs appear to have come to Okoboji in state hatchery trucks. Maybe they would have made it here eventually, anyway. But we will never know this. We are stuck with them now and there is no good news in this. If you would like to help Okoboji's native amphibians, hunt bullfrogs. But make sure that what

you're killing are bullfrogs, not leopard frogs. The bullfrog call is an unmis-
takable BA-ROOM. Bullfrogs lack spots on their back, although their limbs
may be spotted or barred with darker green.

Once again, portions of the following description are drawn from Mary
Dickerson (1908).

Bullfrogs are a more uniform green color than any of Okoboji's other
frogs. Arms and legs may be spotted or barred with darker green. Underparts are white and are variously mottled with dark. The throat of the male
may be yellow. In the eye, the iris is either golden or reddish brown.

Bullfrogs are the largest of Okoboji's amphibians, both sexes often reaching a length of seven to eight inches. Along the back, lateral folds are lacking. A prominent fold extends from behind the eye to the arm, curving
around the back of the ear. Toes are broadly webbed; no joints free, except
the last of the fourth toe.

The bullfrog is more thoroughly aquatic than any of the other frogs of
northeastern North America. They prefer large ponds or lakes and are relatively intolerant of drought. Bullfrogs are late in coming permanently from
their hibernation. It may not be until June that we hear them or see them
sitting along the shoreline or floating among waterweeds at the surface.

Bullfrog tadpoles overwinter and so, alone among Okoboji's amphibians,
can be found any month of the year. They are big—as long as and perhaps
longer than four inches. But they are not noticeably larger than the largest
leopard frog tadpoles in Okoboji. Therefore size is not the best indicator of
their identity. Bullfrog tadpoles are almost uniformly light green and have
small black spots that look like poppy seeds. Leopard frog tadpoles are dark
green to brown above and on their sides, with an abrupt shift to a lighter
belly. The two species will not usually occur together except during the earliest stages when bullfrogs invade a particular habitat.

A bullfrog may sit for hours without movement but fully alert. If a
dragonfly skims over the water's surface in front of him, the insect is taken
quickly. If a sparrow comes for a bath beside what seems to be a moss-
covered stone but is actually an adult bullfrog, its brown tail feathers are
seen a moment later protruding from the frog's mouth, while the frog sits
sedately in just the same spot.

MUDPUPPIES

Mudpuppy larvae and adults, being salamanders, superficially resemble tiger salamander larvae. They are elongate, have four limbs with four toes on each, a tail with a fin, and external gills. They can get big—twice as big as tiger salamander larvae. They are permanently aquatic and tend to live in big water systems—lakes and rivers. How do mudpuppies avoid fish predation when other amphibians cannot? This is a good question. Being secretive and nocturnal must help.

Blanchard reports sightings of one, possibly two, mudpuppies in 1920. They may never have been numerous in Okoboji. It is possible that they still exist. Sampling the bottoms of lakes, especially when they are choked with plants, is not easy. Nevertheless, in five summers of seining the shallows of Little Miller's Bay, we have never caught one; nor have we seen any when searching with scuba gear; nor have we caught any in our traps. Mudpuppies appear to be declining across much of their range. Our data suggest that this trend includes Okoboji. In Okoboji, mudpuppies were last seen in Emerson's Bay in the mid-1960s. Portions of the following description are drawn from Sherman C. Bishop (1948).

Mudpuppies are found under the most diverse aquatic conditions. They may be abundant in the clear waters of lakes and streams but also occur in muddy and weed-choked drainage ditches.

Mudpuppies attain a maximum length of about seventeen inches. Their eyes are small; legs are short and stout and have four toes on both hands and feet. Their color varies but in general is a rust-brown base color, with rounded spots of blue-black scattered over the back and sides. The upper surface of the legs is usually spotted, and the margins of the tail are often tinged with orange or red. The belly is usually light.

Mudpuppies mate in the fall. The female stores the sperm and fertilizes her eggs prior to depositing them in late May or early June. Unlike most amphibians, she remains with them through the period of incubation, perhaps longer. The eggs, attached singly to the lower surface of a shelter, are light yellow spheres about one quarter of an inch in diameter. The eggs hatch in thirty-eight to sixty-three days, depending on the temperature of the water.

Newly hatched larvae are a little less than one inch in length. The color pattern of the larva differs markedly from that of the adult. In particular, the larva has a strongly pigmented median dorsal stripe that originates on the snout and extends the length of the trunk and on the tail, fading toward the tip. This stripe is bordered on each side by a narrower yellow band extending from the gills to the basal third of the tail. Below the lateral light bands is a dark band. The lower sides are somewhat lighter and fade to the yellow, yolk-distended belly. Sexual maturity is attained in about five years at a length of about eight inches.

MISSING PIECES 2

As mentioned previously, when we repeated Frank Blanchard's 1920 amphibian survey of Okoboji during the summers of 1991 and 1992, we found that the amphibian fauna had changed. In particular we found amphibians in decreasing order of abundance: leopard frogs, found in twenty-four of thirty-four sites sampled; American toads and chorus frogs, found in eighteen of thirty-four sites; tiger salamanders, found in thirteen of thirty-two sites; bullfrogs, found in eight of thirty-four sites; Great Plains toads, found in three of thirty-four sites; and gray treefrogs, found in only one of our sample sites.

The relative ranking of species according to their abundance in 1920 compared with 1991/1992 is as follows.

Ranking	1920	1991/1992
1	Leopard Frog	Leopard Frog
2	Tiger Salamander	American Toad
3	American Toad	Chorus Frog
4	Cricket Frog	Tiger Salamander
5	Chorus Frog	Bullfrog
6	Gray Treefrog	Great Plains Toad
7	Mudpuppy	Gray Treefrog

Two species recorded by Blanchard were not found by us. The mudpuppy was not collected. To our knowledge it has not been collected since Blanchard's study. One current Lakeside Lab professor who grew up in Okoboji tells of commonly seeing mudpuppies in Emerson's Bay during

The 1992 Vertebrate Ecology and Evolution class seining a wetland near Iowa Highway 86, two miles north of Iowa Highway 9. The alumni of this class alone consist of one Ph.D., two medical doctors, one high-school teacher, two research technicians, and one U.S. Army captain. They demonstrate the quality of students that attend Lakeside Lab. Photo by Michael Lannoo.

the 1960s. Blanchard's cricket frog was also not found, despite collections in the areas where Blanchard reported their presence. Tweed (1938) wrote: "The cricket frog is common but not abundant in the vicinity of the Iowa Lakeside Laboratory during July and August 1938. It was found on the borders of several small ponds and sloughs, on the shores of the Little Sioux River, on the open shores along [Beck's] canal from Miller's to Emerson's Bay, and on the Silver Lake [Fen] west of Lake Park, Iowa." Although I have visited the University of Michigan's Museum of Zoology and inspected the cricket frogs that Blanchard collected, I have never seen Okoboji's cricket frogs alive.

We did find two new species of Dickinson County amphibians: Great Plains toads and bullfrogs. Five adults of the Great Plains toad have been observed, but we have not yet confirmed the presence of their tadpoles, which would prove that they breed here. As mentioned previously, the bullfrog has been introduced and is now well established along the margins of the lakes and throughout the larger wetlands across the county.

What is particularly disturbing about the disappearances of Okoboji's mudpuppies and cricket frogs is that they disappeared when no one was looking, and as a result we have no clue as to why they are now gone. What is, or was, so wrong about today's environment that it can no longer support species that were here—before thirty years ago—for thousands of years? Should we take this as a sign and be concerned, or should we turn away and act as if nothing is wrong?

For more information on the amphibians of Iowa, see *The Salamanders and Frogs of Iowa*, by Christiansen and Bailey (1991).

THE IOWA LAKESIDE LABORATORY AS A STUDENT SEES IT
Maude Brown, 1910

Okoboji—"Place of Rest." What! A school? The best school ever conducted was held beneath the plane trees where the students walked and talked with the master. Are not the Iowa oaks as picturesque as the sycamores of Athens? And, you alumni, would you give a walk through the deep woods with Macbride or a tramp over the morainic knobs with Calvin for discourse with your heathen philosopher?

"We wish to go to the Lakeside Laboratory," we announced impressively to the pilot of the little double-deck steamer at the pier.

"The other boat, ma'am. Here," he called across, "these folks want to go to the bug-house." We walked aboard the little "Queen" just as she in a most unfeminine voice announced her departure, and we rocked out over the water counted rough that morning as a stiff breeze from the north whipped the tops of the white caps. North and south stretched seven miles of water, while a mile and a half away, east and west, the banks rose in wooded slopes.

"I wonder what sort of aquatic creatures those are!" I remarked tentatively to the pilot as he punched a little pink slip and returned it to me. "They seem rather active for whales."

He looked up the bay. "Oh, motor boat races have started. We'll get a good view. The big one's gaining—there—it's passed them!" And in less time than we are using in the telling, the thing passed in a cloud of spray. Two smaller boats trailed behind, and several more followed in the distance.

"Pretty good clip for the water. Look up to the north end there—sail boat race is on."

I looked and saw a dozen graceful sail boats dipping and turning, forming in line for the start, but the pudgy Queen churned her way steadily westward and our view of the race was cut off by the long red-roofed Manhattan that crouched at the bend of the lake.

We soon entered "Miller's Bay." Ahead of us, topping one of the less densely wooded knolls, appeared the low gray and white buildings of the laboratory. The deep woods along the lake, the prairies stretching off beyond, and the wealth of aquatic plant life through which the Queen now chugged disgustedly, gave promise of a richness of material enough to quicken the pulse of any student of nature, whether that pulse respond to heartbeats ecological, ornithological, entomological or geologic.

As we neared the pier, a fisherman on the sandbar held up an armlong pickerel shining in the sun. A boatman was carefully picking over masses of green stuff which he was bringing up from the bottom and was handling as a miser caresses his gold pieces. Little did I realize as I stepped ashore that I had reached a turning point in my life. That, whereas formerly all unfamiliar green growth, whether on land or in the water, had been dubbed collectively as "moss," I was destined to return to my home prating glibly of *Ceratophyllum, Myriophyllum,* and *Pleurococcus* [the scientific names of plants].

A tall young man who had been gyrating along the surface of the earth at the end of a butterfly net came up and escorted me to the cottage, explaining that, this being Saturday morning, the students were scattered on field work along the lines of their special interests.

He introduced me to the kindly housekeeper, who invited me to rest on the broad, screened-in porch which ran along two sides of the cottage in which were contained the living rooms of the station. She pointed out to me the row of tents at the crest of the hill and explained that these accommodated the men students, while down the path and past the spring was the farm house where the women lived. The laboratory, a low H-shaped structure, stood to the north, and a small cottage behind.

Under a tree lay two high-school boys from a nearby cottage, being tutored in mathematics by an ambitious young man who was paying his own expenses through college.

Soon groups began to return. A half dozen ornithologists with bird glasses swinging over their shoulders were enthusiastic over the newly found nest of young bitterns. Another group laden with baskets talked learnedly of agarics and slime molds, while down the hill came a lone enthusiast boasting of twelve species of grasses from High Point.

The Lakeside Lab boathouse in 1937, one year after it was built by the Civilian Conservation Corps. This building still exists today but is nearly unrecognizable as the bottom level of the lakefront Limnology Lab. On the hill above sits Main Cottage with its wraparound front porch. This is the building Maude Brown described in 1909; it is still in use today. Photo courtesy of the Iowa Lakeside Laboratory.

All gathered on the porch for a few minutes' rest before going on for lunch, and later, while waiting for the mail.

A trim little launch, the "Old Gold," floating an "Iowa" pennant, came up and carried the geology class away to study the natural rip-rapping along the shores, while I went in and duly matriculated as a student in that most delightful school, the Okoboji Lakeside Laboratory, established by the alumni of the State University of Iowa.

The next day being Sunday, we gathered for religious services in the laboratory room, the sermon being the usual, from one of the representative preachers or teachers of the state. After dinner, at which the speaker was the guest of honor, the day was passed quietly, many students gathering about the big library table writing home.

On Monday, the real work began. Classes in the various natural sciences occupied the morning; classes and field work, the afternoon. Often the

Shimek Lab soon after its construction was completed in 1936. Today, this is the lab that faces Iowa Highway 86. During its first years, this building functioned as the library. Today, it is a classroom and houses the vertebrate biology collection, which is composed of many specimens from the nineteenth century that predate the lab itself. Photo courtesy of the Iowa Lakeside Laboratory.

classwork was entirely out-of-doors, the students grouped informally about some object of interest.

About four-thirty, all took the daily dip into the lake and swimming lessons were generously given by those of natatorial accomplishments.

The evenings were variously spent: the collectors over their pins, cyanide bottles and plant presses; the bookish, over their study tables; the musical, around the farm house piano; and some of the undergraduates using the glorious moonlight nights as God intended them to be used. Often the whole family went for a sail after supper, and the indescribable glory of the sunset waters is something that will not be easily forgotten.

Two nights a week, the best talent of the university was drawn upon for illustrated lectures on scientific subjects. These lectures, as also the Sunday services, seemed to fill a long-felt want among the cottagers who attended in weekly increasing numbers.

Altogether, when the all-too-short weeks were over, they were voted entirely profitable: by the high-school teacher and college professor as they nailed up excelsior-stuffed crates of material for the winter's use; by the

The cozy interior of Shimek Lab, when it was a library. No one knows where the elk head came from, or where it went. Elk were common in presettlement Okoboji. Photo courtesy of the Iowa Lakeside Laboratory.

undergraduate with some well-earned "credits"; by the special student, hugging a few more facts developed in the summer's study of his hobby; and by the plain citizen who has learned to love outdoors more through the new wonders revealed. As they once more step aboard the little steamer, there is an intangible something about them that speaks of health—not only of renewed bodily vigor, but of something down deep in their natures that has been renewed and refreshed, and from which life-giving currents will flow throughout the year. So that the teacher will bring truth nearer to the hearts of men, the investigator will see with keener insight, and the citizen will love God and man the better.

6 WILDLIFE MANAGEMENT OR RESOURCE EXPLOITATION?

Although I am not yet ready to concede this point, in the twenty-first century human population pressures will virtually dictate that humans manage most, if not all, ecosystems. What does management mean? Will humans manage for all species, or the most common species, or the most endangered species, or the most economically important species? As the following series of essays will demonstrate, many of Okoboji's wetlands are currently being managed for a short-term economic gain. It is an understatement to say that our wetlands, and in fact aspects of our economy, are not benefiting from this management philosophy.

ABOUT THE AUTHORS

William Barrett appears to have grown up in Okoboji around the turn of the century and was certainly resourceful—an entrepreneur.

Paul Errington was Iowa State University's first world-class wetland ecologist. Errington was a student of Aldo Leopold and on the staff of Iowa State from 1932 until his death in 1962. Errington is the author of several books on wetlands, vertebrates, and natural history, including Of Men and Marshes, The Red Gods Call, Muskrats and Marsh Management, *and* A Question of Values. *When I was a freshman at Iowa State in 1975, my English Composition instructor, who was nearing retirement, wanted to make the point that one could write about anything one wished. "A friend of mine wrote books about muskrats," he said—my personal introduction to Errington.*

Judge C. S. Bradshaw was active in the Okoboji Protective Association and its president from 1928 through at least 1930.

WETLAND IMPORTANCE

Wetlands perform several functions everywhere they occur. Wetlands, without the benefit of added fertilizers, are more productive than our Iowa cornfields. Wetlands that fringe our larger lakes filter dissolved substances—nutrients and contaminants—and settle suspended materials from rainwater runoff. (The title of a recent U.S. Environmental Protection Association Technical Publication is *Created and Natural Wetlands for Controlling Nonpoint Source Pollution.*)

Isolated wetlands drain our fields as they collect rainwater. Fewer wetlands mean more downstream floods. By holding water, wetlands recharge our aquifers and resupply farm fields during drier times. We get our well water from these aquifers. With a reduction in wetlands, aquifers recharge much more slowly, and wells must either be dug deeper to find clean water or clean water must be piped in.

Wetlands benefit wildlife. They are duck factories. As wetlands have been drained in Okoboji and throughout the prairie pothole region, duck numbers have dwindled. Muskrats and mink—commercially important mammals—have suffered from wetland losses, as have song birds, bats, and amphibians, all of which feed on emerging wetland insects, including mosquitoes.

The weir connecting Garlock Slough to West Okoboji's Emerson's Bay. Here, one grate has been removed, allowing both large and small fishes to move into Garlock, which now serves primarily as a spawning ground for carp and bullhead. Photo by Michael Lannoo.

Without these wetlands our recreational lakes become dumping grounds for all the grunge in rainwater runoff. Wetlands serve as sinks for chemicals. Fertilizers that ordinarily wash into wetlands to produce cattails, without wetlands wash into our lakes to produce algal blooms. Farm chemicals and roadside runoff that would be biodegraded in wetlands, without wetlands go directly through culverts into our lakes and contaminate our fish.

An unfortunate example of a compromised wetland is Garlock Slough, which we will consider in more detail later. Before the weir was constructed connecting the north end of this wetland to Emerson's Bay, on West Okoboji, Garlock collected and contained the runoff from the surrounding hills and from Iowa Highway 86. With the weir now providing direct access, this runoff can flow directly into West Lake. For an example of a healthy wetland, consider the Gull Point Marsh complex, which successfully filters a portion of the surface runoff from Wahpeton before this water has a chance to enter West Okoboji.

The wetlands fringing our lakes protect them, yet these are often the most vulnerable wetlands to development. Why have a "wasteland" when instead—with fill—you can have a lakefront condominium complex and a big dock? Why have a "wasteland" when you can dig a canal, put in a weir, and have a "fishery"? Wetlands are natural cleansers, quietly maintaining the water quality of our lakes. Yet because they work so subtly, they are underappreciated, and as such become prime candidates for destruction. Furthermore, like our automobile example in chapter 1, wetlands do not function optimally without all their components in place. You cannot replace wetland amphibians with game fish and expect a wetland to thrive. Without its natural components the wetland filter breaks down.

WETLANDS AND LAKES AS DISTINCT ECOSYSTEMS

In Okoboji, fish and amphibians occupy separate habitats. Fish live in lakes, amphibians live in wetlands. Fish are the top aquatic predators of lakes; tiger salamanders, the top aquatic predators of wetlands. Fish feed during the day and use visual cues to take prey; tiger salamanders feed at night and use nonvisual cues, including their mechano- and electroreceptors.

How do prey respond to predators that detect them using vision? They evolve to become visually inconspicuous, developing a small size and transparent bodies. How do prey respond to mechanosensory and electrosensory predators? This remains an open question. One way would be to move less—to avoid the mechanoreceptors, but how does an animal become electrically inconspicuous? This is unknown. Every time a heart beats or a muscle twitches, an electrical current is produced.

One trend holds. Run a plankton net through an Okoboji lake and you find small, transparent *Daphnia*—visually inconspicuous. Run the same net through a natural Okoboji wetland and you find huge *Daphnia* colored red with hemoglobin—the protein of blood. There is a scientific debate about whether hemoglobin functions in wetland *Daphnia* by allowing them to retain oxygen (remember that wetlands become hypoxic at night) or whether hemoglobin pigment protects these tiny animals from the damaging effects of the sun's ultraviolet radiation. Big red *Daphnia* cannot exist in wetlands containing fish—they are visually conspicuous and defenseless. When fish are introduced, big red *Daphnia* become the first prey taken. Once these zooplankton populations crash, the entire food web—i.e., the

Sketches of tiger salamander larvae in position to feed on mosquito larvae.

ecology—of the wetland is threatened. If zooplankton, the predominant prey of many aquatic animals, are gone, what will these other animals eat? And if these animals are gone, what will the larger animals that normally feed upon them eat?

MOSQUITOES

Wetlands get blamed for the summer mosquito problem. The historical solution has been to drain wetlands. And it's true that mosquitoes breed in wetlands, but mosquitoes also breed in bent rain gutters, old tires used as boat guards along docks, upright wheelbarrows, discarded pop cans, tree holes, and so on. Before we begin blaming wetlands for our mosquito problem, we should eliminate these other sources of mosquito breeding habitat. In fact, healthy wetlands that support tiger salamander larvae, as well as insects such as dragonfly and damselfly larvae, will usually not produce many mosquitoes. These predators will eat large numbers of mosquito larvae. In observation tanks, mosquito larvae—wrigglers—elicit a quick and effective feeding reaction from our salamander larvae. We have found that a single tiger salamander larva will eat about three hundred mosquito larvae per hour—an average of one every twelve seconds. Furthermore, adult amphibians and odonates will eat adult mosquitoes. The mosquito problem in Okoboji can be partly attributed to the fact that many of our wetlands no longer support mosquito predators. In 1920, Frank Blanchard found that every wetland that was large enough to support tiger salamander larvae had them. Recently, we found tiger salamander larvae in fewer than one half (thirteen out of thirty-two) of the wetlands that could support them. Again, wetlands without tiger salamander larvae will produce more mosquitoes than those with them. An irony of chemical pest control is that, by affecting larger predatory organisms, these chemicals may be eliminating our sources of natural pest control.

CONSERVATION
Thomas H. Macbride, 1911

Conservation means the wise use of any utility. The essential idea is use, but use combined with intelligence. It differs accordingly from preservation, which does not imply use at all. If we preserve anything, we keep it indefinitely; if we conserve it, we keep it and use it.

Every bit of practical conservation accordingly implies use. When we advocate the conservation of a lake, therefore, we do not mean simply that we would have a body of water occupying so much area on the ground, but we urge that wherever such body of water of convenient size and depth occurs it shall be kept, and *kept in order*, and *used*. It shall be open to all the people for their use and benefit, for their enjoyment.

FROGGING IN IOWA 1
William Barrett, 1964

A fellow came to Spirit Lake in about 1901 or 1902 by the name of Lu Shumaker. He was the one who gave us the idea of frogging. He hired us to work for him and naturally, it wasn't long before we found out that we could do the work for ourselves and make more money out of it. We frogged in the summer months, starting in the spring.

When the ice was out, the [overwintering adult] frogs would drift in with the waves and we would go out with hip boots which didn't do us much good when we were wading around with the waves washing into us. As the frogs would drift in toward the shore, we would just pick them up. We would work all day and get so cold we couldn't stand in the water. In those days we used to get 10 c[ents] or 12 c[ents] a dozen for them.

FROGS
Anonymous, 1906

Something must be done to protect the frogs. Every year they become scarcer and scarcer.

. . . If you were told that ten million (10,000,000) [leopard] frogs were shipped by only three men from Okoboji last winter, you would be astonished. Yet we have reliable information that this is so. Probably as many more were shipped by other parties. The first day of May, one man caught fifty-four dozen which were full of spawn and were sold for eight cents a dozen.

Have you noticed any increase of the pest the mosquito, and of other obnoxious insects? Yes! Well this is due to such an enormous decrease of frogs.

. . . We must secure a closed season for frogs and stop their shipment to market.

FROGGING IN IOWA 2
William Barrett, 1964

The frogs would spread out through the country when the warm weather came and would stay under the rocks in the spring and under the leaves in the summer for protection. In the summer we worked the fields and picked up one here and there, but we got good prices for them because they were then scarce. We got around 20 c[ents] a dozen in the summer months but never found many of them because we had to walk continuously and scare them up. In those days a man earned wages of around $1.25 a day and we were making from $2.00 to $3.00 a day frogging, so we made good salaries for those times and had easier work too because we didn't have to work every day.

The latter part of the summer, the frogs would go to the fields and meadows. After the fall came on, they would work to the ravines and lowlands and go down the ravines towards the lake. When they started that, we would catch more frogs because they would be in bigger bunches and we would make more money.

We used to ship to commercial men in Chicago, Philadelphia, St. Louis and Minneapolis. Those were the only places where I shipped. Later in the fall I rode around to several of the big hotels, one in Davenport, the Chamberlain in Des Moines, etc., and also shipped to a hotel in Rochester. I would have standing orders of so many per week, say 40 to 75 dozen, and I got about 25 c[ents] to 30 c[ents] a dozen for those orders. I had about eight or nine hotels. One in Dubuque took 50 dozen a week.

One day Claude Farmer and I frogged out at Indian Lake about nine miles northwest of Spirit Lake with a horse and a light wagon. We got so many frogs that we had to walk in and lead the horses. We had no place to ride there were so many frogs. We had them stacked up in one sack on the horse's back. That catch made me $37.50 that day and that was a lot of money in those days for young fellows.

As the days kept getting colder, these frogs went right up against the lake and lay under rocks on the shore so they would be ready to go in when it did freeze up. The day before it would freeze there would be a blanket of them going in.

The frogs liked to be in sloughs. In Indian Lake there was a slough and a sand bar. One day, just before a blizzard came, there were so many frogs going over the sand bar you couldn't see the sand. It was just one mass of frogs. Just a blanket of them seemed to go out into the lake and the lake froze up that night.

After the lake got froze up and before the ice got real thick we would chop holes in the ice to find frogs. We would throw a canvas over our heads so we could see down the bottom. Every once in a while we would find frogs stacked up like a shock of grain or a bee hive. We had a fish pole and we would put fish hooks on it and make spears. We would have probably 15 hooks and jam them down in the ice and shake them out. Just jab and shake—that was the way we got them. Then we would have to hunt again until we found another pile of thousands of them stacked on the bottom. The water was probably 4 feet deep where we caught them. The hooks usually pierced their legs, not killing them.

Trapping frogs was different. Trappers would make fences of chicken wire about 3 feet high which they would stretch along the shore of a sand bar. They would go back almost a quarter mile in a "V" shape. They used chicken wire covered with tar paper so the frogs could not jump through it. Then the frogs would get down in the "V" shape and we could catch them and put them into sacks. They would keep jumping right down in the "V" and then you could scoop them up.

When I frogged we didn't take the small frogs. We had to take the ones of uniform size. Later they took everything. We had to clean our frogs and they had to be dressed. Back from the lake we dug pits—regular cellars— and put the frogs in there. It had to be fast work and after the lake froze up we had to take care of these frogs and transfer them to cellars, granaries, barns—anyplace we could find. We went right to work and dressed them for the market. Pits were quick storage so that we didn't have to monkey with them then and after the lake froze up we would cover up these pits so the frogs wouldn't freeze. Some of the boys left them in the pits and dressed them from there.

To dress a frog, you cut them off at the legs, skin the legs with a knife and pull the skin right off the legs like you would pull off a glove. The hind legs were the only parts used. In later years they sold the whole frog. We had live crates. One crate would have eight or ten shelves in it, but you couldn't put the frogs in too thick because they would smother if you did.

In frogging we never used fences. We did it a little differently. We dug pits and dumped the frogs into the pits. We dug a kind of ditch along the pit to keep them from getting into the lake. Later they improved on that method by getting the fences. Our first idea was to dig pits for them to fall into.

We used to go to a lake where there was a spring. After the frogs went into the lake, they found a place near the spring. It kept the ice open longer and there was more air in the water. You will always find frogs in the springy part of the lake. We went up to Lake Park. The west shore had springs and that was where we made our best catches with the spear after the water froze.

I sold one fall to a company in Minneapolis I had a contract with. I sold them from the first of July until the season was over at 8 c[ents] a dozen.

They bought my stuff straight through. I got more that way than I did otherwise because the price was steady. At times the price dropped to 3 c[ents] or 4 c[ents] a dozen, but later they would be up to 30 c[ents] a dozen.

Thousands of dollars of frog checks were cashed at the banks of Spirit Lake. The whole town was frogging in those days. Men went and even children because they could make such good money at it, but I don't believe the frogs will be back again in such large numbers. The sloughs have been drained and the ground broken up.

AMPHIBIAN DECREASES AND THE DECLINING AMPHIBIAN POPULATIONS TASK FORCE

Biologists have become alarmed at a disturbing, global trend of amphibian decreases. We might expect decreases to occur in species or populations directly exposed to human influences (for example, housing developments and mall projects). However, declines are occurring in populations far removed from direct human influence, in regions set aside as preserves and once thought to be pristine. An alarm call has been raised, and biologists have begun examining the problem.

The first question is, are amphibians declining? This turns out to be a difficult question to answer because amphibian population sizes fluctuate with environmental variables, such as drought cycles. Furthermore, to document declines one needs historical data, which are not available for most regions. As I have shown, Okoboji, with its Lakeside Lab, is special in this regard. Established near the turn of the century, Lakeside Lab has been able to provide historical data on amphibian abundance for Dickinson County. And here, yes, amphibians have declined.

The second question is, what is the cause of these amphibian declines? In addition to wetland degradation, in the eastern United States acid rain falling on unbuffered soils and in unbuffered waters reduces pH levels. These soils and waters are naturally acidic, due to organic decomposition and a lack of naturally occurring calcium (lime) in surrounding bedrock or surficial materials. Add to this the acid rain from sulfur dioxide emissions and you get pH levels similar to those found in cans of soft drinks. Remember the grade school science project, where you put an iron nail into a can of your favorite cola, and when you checked a week later there wasn't as much nail as there was before? Many amphibian embryos are now raised in waters with pH values at similarly low levels. Amphibians are not as tough as nails. Metabolic activities involved in amphibian development are pH dependent. Low pH levels produce a large number of developmental deformities and cause a high mortality among embryonic and larval amphibians. So far, the effects of acid rain in the United States have been felt most strongly in the Northeast, where the prevailing winds carry the airborne effluent of the industrial Midwest.

High-altitude amphibian populations are suffering from high rates of ultraviolet (UV)-B light exposure, which is stronger where the air is thinner. The cause appears to be ozone depletion. UV light is a mutagen that in humans causes skin cancer. Under intense UV light conditions, as is the case in high altitudes, amphibian mortality is high.

In Iowa, acid rain and UV-B light penetration do not appear to be serious environmental problems; not yet anyway (although an UV index is now included as a portion of the summer weather report). Current environmental concerns revolve around agricultural chemicals. Amphibians suffer a double whammy from these chemicals. First, amphibians are directly exposed to pesticides in their environment. Second, pesticides destroy not only agricultural pests but also the insects that provide the amphibian's food—which, by the way, include agricultural pests.

A more recent chemical threat comes from the growing application of lawn-care chemicals. Pesticides sprayed on lakeshore property before a rain end up washing into the lake and killing the plants and aquatic insects. Those inconspicuous caution signs that lawn-care companies put in your yard after they spray are there for a reason. These chemicals are more dangerous than the companies would have you believe. Otherwise, why bother with the signs? The signs are never large enough to provide advertising. We are now learning that children may be especially susceptible to these chemicals. Do you ever wonder why there are so many links between the health of our children and the health of our natural history?

The World Conservation Union, as part of its Species Survival Commission, has established the Declining Amphibians Populations Task Force (DAPTF). The DAPTF has been formed to determine the extent of amphibian declines and to analyze the causes underlying them, and has over

3,000 members and divisions around the world. Iowa is in the United States Central Division, along with Missouri, Illinois, Indiana, and Ohio. For a history and documentation of amphibian declines, I refer readers to Kathryn Phillips's *Tracking the Vanishing Frogs* (1994).

PRESERVATION
Paul Errington, 1987

Feeling as I do, it is hard for me to understand the willingness of the public to drain marshes even if the land so drained might produce corn or some other profitable agricultural crop. Monetary profit should not be the sole objective for land use. We need cornfields and economic bases for our civilization, but we also need marshes where they may be said to belong.

RAISING GAME FISH 1
Judge C. S. Bradshaw, 1930

We found that the Fish and Game Department was putting millions of [walleye] pike fry into our lakes, shortly after they were hatched, but the results did not seem satisfactory. A very small percentage of this crop seemed to survive the rigors of the first few months, and it was suggested by some Iowa sportsmen, notably Dr. Boone of Ottumwa, that nurseries be established in which the pike fry could be protected during the first year and transferred to the larger lakes when old enough to care for themselves. Our Association [the OPA] got vigorously behind this program, with the result that in 1927 Center Lake was set aside as a nursery and was seined of other fish, the best that could be done with the equipment at hand, and many millions of pike fry put into it. In the spring of 1928 the result was very encouraging and approximately one and one-quarter million of fingerling pike, about four inches long, were transferred from Center Lake to West Okoboji. Fishermen report that they were very much in evidence in 1929 and ought to be more so this year.

Huge Department of Natural Resources nets, as well as other supplies and equipment, on a barge in preparation for seining Welch Lake for gamefish fry. By this time of the year (mid-to-late June or so) the herbicide Aquazine has been applied. This biocide kills the submerged aquatic plants and is used to facilitate the efficiency of this seining operation. Photo by Michael Lannoo.

The process, however, was experimental. The Fish and Game Department found that experiments of that kind in other states and by the United States Government had not been satisfactory. That department, however, is giving it a very fair trial. In 1928, Center Lake was restocked with fry, but the equipment at hand was insufficient to thoroughly clean the lake of larger pike put in the year before and the result in the spring of 1929 was disappointing. It was found that the 1928 crop had disappeared. It is believed that the larger pike devoured them. They may have perished for lack of sufficient food. At any rate, the Fish and Game Department, under the vigorous direction of Mr. Albert, has this year purchased a line of new equipment sufficient they think to thoroughly seine Center Lake, and will renew the experiment. They have also set aside Diamond Lake and Welch Lake as nurseries, which were stocked last year with fry. The Department has even been pumping oxygen into Diamond Lake during the winter season.

RAISING GAME FISH 2
Anonymous, 1929

Another improvement at the upper end of the lake territory has been the setting aside of another lake as a pike nursery. This will be at Welch Lake, five miles northwest of town, on the Reinen farm. It will be cleared of bullheads before the small pike are put into the water.

As a result of the fish hatchery, which the OPA was instrumental in getting for the lakes, there have been added millions of fingerling pike in all of the Dickinson County lakes. The small fry are first put into, and left for a year in, Center Lake, before removal to the larger lakes.

Through the Fish and Game Department a black bass hatchery has been established in the canal between Miller's and Emerson's Bays. All other fish have been removed, by seining. The black bass are encouraged to spawn in the canal, and the small fry are protected.

A number of other lakes, including Welch Lake and several other of the smaller lakes, have been set aside for refuges for pike fry, until the fingerling stage has been reached, after which they are transferred to the larger lakes.

Several hundred thousand lake trout have been placed in West Lake Okoboji, which is the only lake in the state suitable for these fish. Fingerling black bass from the hatching ponds of eastern Iowa have also been placed in our lakes [as have been] a number of adult black bass.

A MISSING ZOOPLANKTER
Kenneth Lang

The tiny zooplankton, *Daphnia longiremis,* is no longer found in Okoboji. First discovered in 1920, this animal, a portion of the prey base for the fish in West Lake Okoboji, has not been seen since. This species prefers cold temperatures and is usually found to the north, in the lakes of Canada and the northern United States. In Okoboji, in 1920, it was found in the deepest and coldest part of West Lake. By 1939, fewer than two decades after it was first discovered here, it was gone. Why then this disappearance? Two factors may be responsible. First, the lake has been eutrophied—the byproducts of human use have led to an increase in the amount of organic

The weir connecting Hale Slough and Big Spirit Lake. Note that while the grates keep out larger fishes, smaller fishes, including the fry of larger fishes, are free to swim into the wetland. Photo by Michael Lannoo.

matter dumped into the lake. During the 1920s and 1930s, Okoboji's human population grew, agricultural practices were intensified, swamp busting programs were enacted, and at that time the sewage system had not yet been built. While many of these practices no longer continue, eutrophication does. This eutrophication adds, each year, about fifty percent more organic matter into the lake than was added during presettlement times. Organic matter begets organisms, and organisms use oxygen. As a result, since 1920, there has been a decrease in the amount of oxygen available towards the bottom of the lake, where *Daphnia longiremis* was found. This species may now be gone because there is no longer enough oxygen available to sustain it.

A second possible explanation is that *Daphnia longiremis* was the victim of a fish species introduction. Nonnative lake trout were put into West Lake during the 1920s. Lake trout also like deep, cold waters and would have been found at the same depth as *Daphnia longiremis*. Furthermore, lake trout are zooplanktivorous—they eat *Daphnia*. Did *Daphnia longiremis* succumb to lake trout predation? We will never know. The lake trout themselves have also become locally extinct. Okoboji was no place for them.

TOP PREDATORS

Why aren't tiger salamander larvae found in lakes? In Okoboji, fish and lar-val amphibians sort themselves by habitat type. Fish tend not to be found in wetlands because wetlands either dry up or become anoxic. Fish do not have lungs like amphibian larvae.

Amphibians aren't found in our lakes because young larvae are small and established fish are large. Fish will eat the amphibian larvae. If you doubt this, I suggest looking up the 1986 *In-Fisherman* magazine article by Carl-son and Stange, "Frogs and Waterdogs—The 'Right-Stuff' for Big Bass, Pike and Walleyes." This reeks of macho, and because of this attitude there is a tendency to think that in this predator-prey relationship the fish are al-ways the predators. This is incorrect. Fish are established in lakes—they live there as adults. They are large. Aquatic amphibians must start all over each year from eggs. Young amphibian larvae are small. The bigger fish eat the smaller amphibians. When larval hatchery fish are introduced into wetlands with established salamander larvae, the tables are turned and the larger sala-manders gorge themselves on the larval walleye and muskies. From a fish manager's standpoint, this is not a cost-efficient operation. Therefore, it be-comes necessary to tip the scales (so to speak) in favor of the game fish lar-vae by eliminating the salamander larvae. This is now done in Okoboji, in the spring, by using a chemical called rotenone.

ROTENONE AND AQUAZINE: STATE-APPLIED AQUATIC TOXINS

According to the DNR publication *Iowa Fish and Fishing*, by Harlan and Speaker, rotenone is a fish toxicant used reluctantly, only when the "fish population structure [is] completely out of control." This is not true. Rotenone is used more often than this and kills more than fish, as the data from studies conducted by H. L. Hamilton (1939) at the Iowa Lakeside Lab in the late 1930s show:

"The powdered root of *Derris elliptica*, which contains the alkaloid, ro-tenone, has long been used by the natives of tropical countries as a fish and arrow poison. Recently it has been introduced in this country as an insecticide, and is being used by [managers] to remove rough fish from waters which are to be stocked with game fish. Little or no consideration has been given to the possibility that the poison might kill other members of a lake fauna (thus breaking the food chain) and seriously affect the survival of any fish with which the waters might be restocked. The purpose of the present investigation was to determine the effect of rotenone on various aquatic animals.

"The paralyzing action of rotenone on the respiratory center in mammals is well known. Fish respond to the drug by swimming to the surface and gasping for air; the amount of air swallowed is soon sufficient to float them ventral side uppermost, and the animals writhe on the surface of the water in this position until death results from suffocation. The lethal concentration (that which produced death in at least twenty-four hours) for various animals is given below [partial list].

Animal	Concentration
Buffalo (*Ictiobus* sp.)	1:6,000,000
Carp (*Cyprinus carpio*)	1:5,000,000
Bullhead (*Ameiurus melas*)	1:2,000,000
Leptodora kindti (zooplankter)	1:2,000,000
Diaptomus (zooplankter)	1:2,000,000
Daphnia (zooplankter)	1:2,000,000
Leopard Frog (*Rana pipiens*)	1:500,000
Tiger Salamander (*Ambystoma tigrinum*)	1:500,000

". . . Gill breathing animals were more sensitive to rotenone than air breathers. A concentration of 1:500,000 was lethal to tadpoles of *Rana pipiens* within eight hours, but metamorphosed animals could tolerate the same concentration for 24 hours. In general, the more exacting the oxygen requirements of the fish, the more sensitive it was to rotenone. It was found that rotenone was rapidly decomposed in water from West Lake Okoboji and that a solution as strong as 1:100,000 was apparently harmless after 48 hours.

"A gross examination of the gills of normal and derris-treated [rotenoned] fish showed a marked difference in blood supply. The gills of poisoned fish were pale pink instead of bright red, and this indicates that

Department of Natural Resources sign warning of Aquazine application at Sunken Lake. The signs are placed around Welch and Sunken Lakes. Because Aquazine persists in these waters for up to a year after application, and because Aquazine has been applied nearly every year, these signs are constant features of the shorelines of these basins. We know how Aquazine has affected the algal flora of these lakes; we wonder how it has affected the groundwater. Photo by Michael Lannoo.

suffocation was due to decreased circulation of blood through the gill filaments. The beat of the heart remained strong, and it is probable that death was due to a vaso-constrictor action of the alkaloid . . ."

In today's Okoboji, publicly owned Sunken Lake and Welch Lake are routinely rotenoned and used to raise game fish—walleyes and muskies—from larvae to fingerlings. This solves the first problem with raising game fish in wetlands—namely, preventing them from being eaten by larger fishes or by tiger salamander larvae.

The second problem with raising game fish in wetlands is getting them out once the wetlands have a mature growth of submergent plants. The current answer is to apply a second chemical—the herbicide Aquazine—which kills the plants and enables the big DNR nets to be run through the newly opened water.

Aquazine is a broad-spectrum, nonselective aquatic herbicide/algicide for use in waters such as wetlands with no outflow. After application, water cannot be used for livestock, irrigation, sprinkling, and domestic use for twelve months—a full year. Application rates: seven-and-a-half pounds per acre foot for weeds. Commercial price: about four hundred and fifty dollars per fifty pounds. Not only is Aquazine expensive, but for a wetland that is earmarked for game fish production every year, it is permanent. Just when a wetland is beginning to recover from last year's treatment, it gets hit again. It is likely that Welch Lake and Sunken Lake have been continuously contaminated with Aquazine from at least 1980, the first year that I began to take notice, through 1992. In 1993, Welch Lake was apparently being given a reprieve from Aquazine, although game fish were raised there. Sunken Lake was poisoned again. One result is a nearly absolute elimination of the algal and submergent vascular plant species of these lakes.

SILENT SPRING CHORUSES

In her book *Silent Spring*, Rachel Carson warned against the dangers to both nature and humans of fisheries management practices similar to those being used in Okoboji. She writes: "In the entire water-pollution problem, there is probably nothing more disturbing than the threat of widespread contamination of ground water. It is not possible to add pesticides to water anywhere without threatening the purity of water everywhere. Seldom if ever does Nature operate in closed and separate compartments, and she has not done so in distributing the earth's water supply. . . ."

She continues: "It is an extraordinary fact that the deliberate introduction of poisons into a reservoir is becoming a fairly common practice. The purpose is usually to promote recreational uses, even though the water must then be treated at some expense to make it fit for its intended use as drinking water. When sportsmen of an area want to "improve" fishing in a reservoir, they prevail on authorities to dump quantities of poison [i.e., rotenone] into it to kill the undesired fish, which are then replaced with hatchery fish more suited to the sportsmen's taste. The procedure has a strange Alice-in-Wonderland quality. . . . yet the community, probably unconsulted about the sportsmen's project, is forced either to drink water containing poisonous residues or to pay out tax money for treatment of the water to remove the poisons—treatments that are by no means foolproof."

And finally: "As ground and surface waters are contaminated with pesticides and other chemicals, there is a danger that not only poisonous but also cancer-producing substances are being introduced into public water supplies." For more recent descriptions of environmental toxins and their dangers, I refer readers to Lave and Upton's *Toxic Chemicals, Health, and the Environment* and Lappé's *Chemical Deception*.

NATIONAL POLLUTANT DISCHARGE ELIMINATION SYSTEM PERMITS

It is fair to ask, as I have, why the pollution of surface waters by aquazine and rotenone is not illegal. Why do we have a Clean Water Act if it does not protect us from this sort of pesticide application? It turns out that a National Pollutant Discharge Elimination System (NPDES) permit is required for the discharge of any pollutant from a point source. Discharges into aquaculture projects are subject to the NPDES permit program. This applies to any managed water area that uses discharges of pollutants into a designated area for the maintenance and production of harvestable freshwater plants or animals. Can a NPDES permit be issued for aquazine and rotenone application? It sure can, and here is why. Iowa is under the jurisdiction of the Environmental Protection Agency's Region 7, in Kansas City, which since 1978 has given the Iowa DNR the authority to administer and enforce the Clean Water Act, and therefore to oversee the issuance of NPDES permits in Iowa. Do you think that the DNR would turn itself down for a NPDES permit? Me neither. There may be an issue of public trust that is being violated here, but as long as Iowa retains the right to issue NPDES permits to itself, it appears that it is perfectly legal for Fisheries DNR to poison Okoboji wetlands.

BULLFROG INTRODUCTIONS

State fisheries biologists have introduced and harbored bullfrogs. (As the implications of this action have become apparent, these charges have been denied. However, through photographs and past interviews a clear and unambiguous picture has emerged of the history of recent Okoboji bullfrog introductions.) Bullfrogs are native to southern Iowa and historically were absent from Okoboji. They are not native to our region. Frank Blanchard found no bullfrogs after five weeks of extensive collecting in 1920. Bullfrogs were successfully introduced according to Erickson (1984) as follows: "According to one employee of the Spirit Lake Fish Hatchery, the hatchery started bringing bullfrogs to the Spirit Lake Area [from the Rathbun area, as a genetic analysis of these frogs would show] in the mid-1960s. For the first few years no frogs seemed to survive the winter. In 1968 or 1969 bullfrogs were able to adapt to the new surroundings and a breeding population was started. The frogs were originally introduced to the pools on the hatchery grounds and in a few years were reported in ponds around the fish hatchery area. The reason given for introducing bullfrogs to this area was to have a new species in the area." (I wonder if DNR fisheries biologists knew which species were here in the first place.) During the summer of 1977, I found bullfrogs to be present only in the Orleans Fish Hatchery ponds and a wetland immediately to the west, across State Road 276. Bullfrogs have increased their range during wet summers and are now found across the county and into Minnesota (Oldfield and Moriarty 1994), a fact the Minnesota DNR is well aware of and not thrilled about. In 1992 bullfrog adults invaded the Gull Point wetlands for the first time, but a winterkill in 1993–1994 knocked back their numbers. In 1995 they returned.

This bullfrog introduction, like most species introductions, was not based on sound ecological principles. To demonstrate the effects that bullfrogs have on our native amphibians, place an adult bullfrog in an aquarium along with any of our native amphibian species. You will find that the bullfrog grows at the natives' expense. In 1993 we determined that an adult bullfrog will eat, on average, about two leopard frogs every three days. We must ask, does the whim of a fisheries manager justify this ecological damage?

Here are some comments from recent scientific papers on the effects of bullfrog introductions:

". . . through larval interactions, bullfrog invasion can perturb aquatic community structure and exert differential effects on native frogs" (Kupferberg 1993).

"The disappearance of *Rana aurora* from the [San Joaquin Valley, California], and the continuing reduction in range of *Rana boylii*, is attributed to habitat alteration coupled with predation and competition from [the bullfrog] *Rana catesbeiana*" (Moyle 1973).

"Successful reproduction of introduced bullfrogs *Rana catesbeiana* in

Signs at the Orleans hatchery grounds protecting introduced bullfrogs. The lower sign, photographed in 1992 and faded with age, proclaims: FROG TAKING PROHIBITED. *Photo by Michael Lannoo.*

northwestern Europe: a potential threat to indigenous amphibians" (Stumpel 1992).

"Concurrent with the establishment and range expansion of the bullfrog in Colorado has been the decline in leopard frogs" (Hammerson 1982).

Bullfrogs are no longer being introduced in Okoboji, but no serious attempt is being made to eradicate them, either. In fact, there seems to be an overall insensitivity to the negative impact that bullfrogs are having on Okoboji wetlands. As I write this, bullfrogs are still being proudly displayed in the live exhibit at the Orleans Fish Hatchery. Bullfrogs are now in so many Okoboji wetlands that eradication efforts would have to be massive, anyway. In Okoboji we have a bullfrog problem. And what are the consequences? Again, by looking elsewhere, we may be able to peer into Okoboji's future. As with Iowa, over the past few decades biologists from the Arizona Game and Fish Department have been spreading bullfrogs across their state. Today, Arizona biologists have a clear choice: control these bullfrogs or risk losing the Chiricahua and the lowland leopard frogs. The Chiricahua leopard frog is a category 1 candidate for federal endangered species listing and will probably be proposed for listing if the Endangered Species Act is renewed. Arizona is finding that controlling bullfrogs is expensive; eradication is out of the question.

Alone among Okoboji amphibians, bullfrog tadpoles must overwinter. This means that bullfrogs tend to concentrate in our larger wetlands—basins that hold water throughout the year, basins that do not dry during droughts, basins that our native amphibians use for successful reproduction during droughts. Aldo Leopold, every biologist's hero, would have objected to bullfrog introductions. He wrote that the continuation of native species is an "ethical" matter transcending any "utilitarian objective" (Flader and Callicott 1991).

POLLUTION
Anonymous, 1915

The lake [West Okoboji] was pretty noisy last summer, owing to the great number of detachable row boat motors in use. It is to be hoped that the makers may design a successful silencer for them.

MUSKIES

In 1967, muskies were first introduced into West Lake Okoboji. Since then they have been introduced almost every year. Muskies are voracious predators. Despite this, their effect on the smaller aquatic vertebrates of our larger lakes has never been documented. Perch fishers argue about the impact of muskie introductions all the time, but did muskie introductions also lead to mudpuppy declines? We will never know, since all we have is evidence for mudpuppy declines at or near the time of the initial muskie introductions—correlative evidence cannot be used to establish cause. Suspicions are aroused, however, when one considers that mudpuppies were probably in Okoboji for thousands of years, were present up to the late 1960s, and have not been seen since. An environmental impact statement that included a biological survey of West Okoboji was not conducted before the introduction of muskies. It would have provided a nice set of baseline data to compare with today's species and their abundances.

THE REFUGIA HYPOTHESIS

Drought and its consequence, the drying of smaller wetlands, is perhaps the harshest natural environmental factor that Okoboji amphibians must face. Droughts may last several years and prevent successful amphibian breeding in shallow wetlands—the majority of Okoboji's wetlands. During droughts, habitats such as Diamond Lake, Welch Lake, Sunken Lake, Hottes Lake, and Marble Lake become functional wetlands and have historically provided—on what ecologists now call the metapopulational level—the opportunity for successful amphibian reproduction. These basins have served as amphibian refugia—buffers against severe population declines during prolonged drought. These habitats are now less than optimal, and perhaps hostile, as game fish have been moved into them. Furthermore, because bullfrogs must overwinter as aquatic tadpoles, these middle-sized basins are their preferred habitats. The result of these fisheries management decisions for our native amphibians is that they are now being affected by drought to a greater extent than has historically been true. In the past, amphibians were forced into refugia by drought, and now they have been moved out of their refugia by aquacultural practices. If they manage to survive, they are twice refugees.

GARLOCK SLOUGH: A CASE STUDY

Garlock Slough is an abused wetland. Once a haven for Okoboji amphibians, it now houses a summer assemblage of carp, bullheads, and bullfrogs, and in some years a small number of native amphibians.

Garlock's demise began thirty years ago, as noted by Bovbjerg (1965): "In our previous report we noted a decrease in the [leopard frog] tadpole population during the years 1962 and 1963. We were at a loss to explain this from any local or meteorological conditions over these years. In 1964 the population was reduced to zero. No tadpoles were found after extensive collecting in the entire slough. Many of the usual invertebrates were in reduced numbers as well. We did, however, find large numbers of fish of several species not seen in previous years. A question put to the officials of the Iowa Conservation Commission at Orleans, Iowa, was answered with the suspected but stunning information that truckloads of young fish were stocked in the slough in 1964, a practice increasing in intensity over the last two years. . . . The wildlife preserve at Garlock Slough is obviously no longer in existence and to that extent at least accounts for the brevity of this report. It had been hoped that a continuing study could have been made of the frog populations and their migratory behavior in what was an ideal natural situation."

Hayes and Jennings (1986) write: "Perhaps the most neglected but potentially important [impact to native frog species] is predation by introduced fishes."

Furthermore, a cement weir has now been built connecting the north end of Garlock to Emerson's Bay. We have mentioned this weir before. One stated purpose of this structure is to increase breeding habitat for game fish—so they will swim into the slough to spawn. In theory, the young fish can then feed and grow on the rich invertebrate food base and be protected from fish predators by the thick submergent plants. When the young get large enough, they can then somehow make their way back into the lake. A second stated purpose of the weir is to trap "rough fish" as they move from the lake to the slough. (Being on the outside, one can only know the results of management policy, not the rationale, which often changes depending on whom one consults, when they are consulted, and the context.)

The problem with the first idea is that Garlock summerkills by early July nearly every year. This is not nearly enough time for game fish young to grow up, find the weir, swim into West Okoboji, and survive there. The problem with the second idea—to trap rough fish—is that they are not being trapped. In fact, the opposite effect occurs. Garlock is contributing to the rough fish "problem" by propagating them. Rough fish—carp and bullheads—can withstand Garlock's summer hypoxic conditions and now have the luxury of Garlock as breeding habitat. We have a problem. The solution? Simple. Why not remove the weir and fill in the hole? Garlock will go hypoxic or dry during the next drought cycle and become a restored

The sign that used to advertise public fishing at Garlock Slough. It has since been replaced by a more accurate sign. Photo by Michael Lannoo.

wetland, amphibian habitat. This wetland restoration will be inexpensive, since the public—you and I—already own the land.

If you doubt the fact that prized game fish avoid Garlock, see for yourself. Run a net through the water and see what kind of fish you find. If you are not so inclined, pay attention to the circumstantial evidence. Until 1993, you may have noticed the large sign at the south entrance advertising "Garlock Slough" and "Public Fishing." Ask this question: When was the last time you noticed someone fishing in Garlock Slough? I, myself, have never witnessed it. This habitat is, and for the next few thousand years will be, a wetland. No amount of concrete and advertisement can make it a lake.

Garlock Slough is an injured wetland, a victim of the aquarium ethic. What we need instead in Okoboji is a "wetland ethic," modeled after Aldo Leopold's celebrated land ethic. Remember Leopold: "A thing is right when it tends to preserve the integrity, stability, and beauty of the biotic community. It is wrong when it tends otherwise." A land ethic holds in Okoboji. We do not see state biologists turning Cayler Prairie into a sod farm to make the state more money—a ridiculous notion, until one considers what is happening to our wetlands. We do not convert our native prairie remnants to agricultural uses, yet we convert our native wetlands to aquacultural uses.

Garlock Slough is currently no more a wetland than a sod farm or a cornfield is a prairie.

And so, you ask, what would a wetland ethic involve? For starters, a consideration of wetland biology, normal wetland function, and a particular wetland's unique history. How should we manage Garlock Slough? To repeat: we should isolate the basin and leave it and its watershed undisturbed. Garlock will once again manage itself the way all wetlands do, through anoxia and drought.

Garlock Slough is unhealthy because it is being viewed by those responsible for its management not from an ethic based on wetland knowledge but rather from a perspective of single species management, of game fish management based on short-term economics. This philosophy is killing Garlock Slough and in the process is injuring our lakes. The citizens of Okoboji could not possibly afford to build the water treatment facilities to duplicate the services being provided by our wetlands. It is time to get Garlock Slough back on line, so that it works to maintain the water quality of West Okoboji.

7 A NEW PERSPECTIVE: A WETLAND ETHIC

Imagine this scene from the next century. Your granddaughter walks down the gravel access ramp to an Okoboji wetland with her young children and, being curious, starts poking around. Remembering how her mother—your daughter— had shown her, she takes a small net, runs it through the water, and kneels to examine what she has caught, to show her children what she was shown. She looks at the net, then looks around, and finally she sees. Her smile drops as she stands, puts her hands on her hips, and gazes out across the quiet cattails to survey what she had first missed. She shakes her head from side to side and whispers to herself, "My God, what happened here?"

This book, while short, has been one long advocacy for the native natural history of Okoboji, and in particular for the conservation of natural wetland habitats and their amphibians. Why bother? Who cares about animals sometimes heard but rarely seen? Who cares about animals with no current commercial value (other than as live bait)? My specific argument, put formally, is that to maintain the health of our lakes, we must maintain the health of our wetlands. And to maintain the health of our wetlands, we must maintain the health of our amphibians. To the layperson, this leads to the seemingly absurd conclusion that to maintain the health of our lakes, we must maintain the health of our amphibians. Yet this is exactly my point: amphibians are the Lakes Area Chamber of Commerce's best friends.

To a biologist this makes sense. For example, today in Okoboji we have a summer insect "problem"—mosquitoes, flies, other annoying and destructive pests. How do we deal with them? By installing bug zappers and spreading pesticides. In the not-so-distant past, insect control was largely the domain of amphibians. How many mosquitoes could twenty million adult leopard frogs, along with a correspondingly high number of our other native amphibians, eat? In the days before pesticides—days to which we are unlikely to return—amphibians were valued for their ability to devour harmful insects. Mary Dickerson (1908) wrote: "That the [American] toad is the gardener's ally has been proved beyond a doubt. The economic value of the toad has been recognized in this country as well as in others. For many years, gardeners in France have been glad to buy toads in order to have them as insect destroyers. . . . The estimated value of a single toad is $19.88 per year [1908 dollars]." (At that time, Thomas Macbride's salary at the University of Iowa was twenty-five hundred dollars.) The economic value of nongame animals is often overlooked. One bat colony in Texas is reported to eat a quarter million pounds of insects each summer night. In Okoboji, we have removed a portion of our natural pest control—amphibians—and replaced it by industrial pest control. Do you think our recreational lakes are healthier for this conversion?

Furthermore, pesticides are undoubtedly contributing to amphibian declines. The process goes as follows. Pesticides are broadly applied and kill most, but not all, pests—some get missed, some get hit but are resistant. Predators—for example, amphibians—eat these dying pests, accumulate the pesticide, and themselves become victims. Pests—smaller, more numerous, faster reproducing—rebound quickly when the numbers of natural predators are reduced, and, of course, must be poisoned again, this time with heavier doses to offset their resistance. Any predators that managed to survive the first application become targets for the second application. Eventually, predators become squeezed out as the natural predator-prey interaction becomes replaced by the pesticide-prey interaction.

A spiral has been started in which natural history components are being replaced by industrial and chemical technology. Where does it end? Gretel Ehrlich, in *Islands, the Universe, Home*, has put these feelings into words: "What shocks me so is the detachment with which we dispense destruction—not just bombs, but blows to the head of the earth, to populations of insects, plants, and animals, and to one another with senseless betrayal—and how the proposed solutions are always mechanistic, as if we could fabricate the health of the planet the way we make a new car." If the Iowa Great Lakes eventually become sterile, who will come to fish our waters? Who will come to shop at the Three Sons clothing store? To enjoy the Okoboji Summer Theater? To ride the *Queen II* excursion boat? To visit the Black Walnut Nutty Bar Stand? Sterility may be an exaggeration, but ask any of the old-timers how the local fishing has changed. Before his death, Cap Kennedy, longtime Okoboji guide, described the huge downturn in Okoboji fishing conditions during his lifetime. Extrapolate this trend into the future. The problem is not with the fish themselves; the hatchery supplies many of these animals. The problem is with the condition of our lakes; our lakes cannot support the aquatic life, including fishes, that they once did.

Some may argue that the past is in the past and that conditions are now fine—that we should stay the course. (At this point, one should be re-reminded of Pearl Harbor and the views of Private Lockard's superiors.) To this, I offer the specific rebuttal that *Okoboji Wetlands* comes at an arbitrary time in the history of our region, not at some watershed event that promises to change the way we think about our natural world. There are no data to suggest that we now deviate from the trajectory noted by Cap Kennedy.

For ecosystems to survive, they need their components intact. I repeat: Okoboji's wetlands need their amphibians. Aldo Leopold has said about ecosystems: "Presumably the greater the losses and alterations, the greater the risk of impairments and disorganizations" (Flader and Callicott 1991).

Some readers may perceive this book as anti–game fish. This, of course, has not been my intent. One value of fishing and hunting is that these activities get people outdoors to learn about nature, to appreciate animals and ecosystems. How many of Okoboji's pioneers came here to fish and hunt? How many of our current biologists became interested in nature through fishing and hunting? My own introduction to biology was through hunting pheasants in northeastern Iowa. Furthermore, the economy of Okoboji hinges on fishing. We need our game fish and our fishers. My complaint comes with how we produce these game fish—poisoning wetlands cannot be justified as a means to any end. Is there any thought given to the other species—for example, the waterfowl, shorebirds, herons, hawks, muskrats, mink, and turtles—that also use the wetland? Is there any thought given to the human families who live across the road or up the hill and take their water from a well that gets recharged by the poisoned wetland? (Shouldn't the cost of aquazine applications include well testing? As I write this, developers are offering lots on Welch Lake.) Is there even any thought given to the game fish themselves? Where in this scenario does respect for nature—or, in fact, human health—enter?

Isn't it time to find a better way to raise our game fish? Isn't it time to be conservative? Although I dislike telling people what to do, I wonder why we do not follow other states in building artificial rearing ponds, in this case near our fish hatchery where land is already available. Artificial ponds would offer managers more control, and therefore more flexibility, in their rearing programs. For example, with rearing ponds, biocides would no longer be needed, drawdowns would control both aquatic predators and submergent plant growth. It is true that building rearing ponds would be expensive, but wouldn't the increased yield in game fish production, and the money no longer spent buying expensive biotoxins, soon offset these costs? Plus an added bonus: the harassment of Sunken Lake and Welch Lake would end, and they could be allowed to revert to their natural ecosystem dynamics.

If further proof of the inadequacy of wetlands to raise game fish is required, I point out the current frequency of failure. Larval game fish do not survive well in wetlands. During good years only a small fraction of fish put in as larvae are harvested as fingerlings. During bad years there is complete failure. When this happens (for example, in Sunken Lake in 1994) fish are purchased from hatcheries in other states. During these bad years, raising game fish must cost the state about twice as much money. Where does this extra money come from?

Poisoning wetlands, bullfrog introductions, weirs connecting wetlands to lakes—none of these state-sponsored activities currently undertaken (a carefully chosen word) on public lands is consistent with a wetland ethic. These activities do not promote wetland (and perhaps human) health; they do not consider the functions of natural wetlands; they do not consider a wetland's unique history. In today's Okoboji, wetlands are viewed by those responsible for their stewardship as aquariums—vessels to raise fish. Therefore, rotenoning a wetland becomes just like purging a fish tank. And bullfrog introductions become an amusement, just like going to a pet store and purchasing a new kind of fish. In fact, to extend the aquarium analogy one step farther, Sunken Lake has electrical hookups and an industrial aerator—an air stone.

This aquarium ethic currently supersedes our wetland ethic. But understand this: compared with natural wetlands, aquariums are poor lake filters; aquariums are sterile; aquariums have little natural history. If you do not believe me, just compare the depauperate algal flora of impacted Sunken Lake (one species confirmed, probably fewer than half a dozen species total) with the diverse algal flora of its more natural neighbor, Hottes Lake (several hundred species present), which lies not more than twenty yards away. Repeat the observation using impacted Welch Lake.

A partial list of Okoboji's impacted wetlands follows:

Pillsbury Lake	Drained
Pratt Lake	Drained
Sylvan Lake	Drained
Center Lake	Historic Game Fish Nursery
Diamond Lake	Historic Game Fish Nursery
Sunken Lake	Current Game Fish Nursery
Welch Lake	Current Game Fish Nursery
Garlock Slough	Weir, Bullfrogs
Hale Slough	Weir
Gull Point Wetland Complex	Bullfrogs
Jemmerson Slough Complex	Bullfrogs
Grover Lake	Bullfrogs

Some may perceive this book as anti-DNR. This has also not been my intent. One questionable management policy does not necessarily define such an institution, and the policies of an institution do not define the quality of its individuals, especially its field people, who must follow policy or be replaced. But we can ask, what happens to our natural resources when those we charge with their protection are remiss? Perhaps the problem lies with how we define natural resources. Does the term natural resource now apply only to fish and game? If so, does this mean that native amphibians must be something else—perhaps natural, but not a resource? Conversely, are exotic muskies and bullfrogs still considered *natural* resources? We often hear that by setting aside habitat for game species we create habitat for nongame species. This may sometimes be true, in fact it may mostly be true, but it is certainly not true in the case of Okoboji's wetlands—the aquarium ethic excludes native species. Is our current philosophy toward natural resources to preserve Iowa's native biodiversity or to make money for state coffers through license revenues? Indeed, these two pursuits need not be mutually exclusive, but, where there is conflict, which goal takes precedence?

Okoboji is not alone here. The trespasses of Okoboji's fish managers have been repeated across the country. In Minnesota, fish aquaculture is a multi-million dollar industry dependent almost exclusively on natural wetlands. Or consider the Green River, a tributary of the once-powerful Colorado in the western United States. "Sponsored by the U.S. Fish and Wildlife Service and implemented by the game and fish managers of the states of Wyoming, Utah, and Colorado, an attempt was made to eradicate by poisoning all native fishes in 500 miles of the Green River and its tributaries" (Carothers and Brown 1991). As Bob Miller (1963) details: "The Green River project was the most extensive eradication job ever undertaken, and since the poison traveled through three states it might be justifiably labeled as interstate pollution that was financed, albeit largely unwittingly, by American citizens. Between September 4 and 8, 1962, more than 20,000 gallons of an emulsified rotenone preparation were applied to nearly 500 miles of this river by more than 100 men. The cost of the poison alone was $157,000. Funds for the project were authorized by Congress in June 1961, after full approval by the Bureau of Sport Fisheries and Wildlife, of the U.S. Fish and Wildlife Service, of the program proposed by Wyoming and Utah."

As the target "trash fish" species, found only in the Colorado and its tributaries, have become endangered, the same agencies responsible for applying rotenone are now charged with recovering these populations. If this scenario sounds familar, it is. It is the Arizona bullfrog story, repeated for fish. Will the story be repeated yet again for Okoboji's amphibians?

In Hawaii, the State Department of Fish and Game has introduced seventy-eight species of birds. A local bird society (Hui Manu) brought in an additional twelve species in just twelve years. One result of these introductions is that today at least one half of the native bird species are extinct; half of those remaining are endangered. Meanwhile, at least fifty species of exotics are thriving (Tisdale 1994).

Pister (1976) has said it best: "We still have a long way to go in the fish and wildlife professions before we reach an acceptable level of philosophical maturity. History alone will judge the value of what we do today. In the year 2076, society will be far less interested in the 1976 catch per angler in Crowley Lake or the degree of hunter success on a certain wildlife management area than in what happened to our native fauna if we fail to appreciate it enough to preserve, manage, and utilize it. We have inherited so much from our predecessors that we automatically assume an enormous debt to the future."

I've drawn on the thoughts and perspectives of Okobojians past— Macbride, Blanchard, Shimek, Hamilton, Thomas, Stephens, past OPA activists, concerned citizens. These were people who knew Okoboji in its innocence and recorded its *natural* history for us to read—our benevolent and wise ghosts. Today I work out of the Iowa Lakeside Laboratory and coordinate the United States Division of the Declining Amphibians Populations Task Force. I am also a member of the Invasive Species Specialist Group (World Conservation Union, Species Survival Commission), where my responsibility is to document the worldwide impact of bullfrog and game fish introductions on native amphibian populations. We at the Lakeside Lab will continue to study the Okoboji amphibians in an attempt to further understand our wetlands and to address the broader problems of global amphibian declines. The goal of Thomas Macbride, Lakeside's founder, to study the native flora and fauna of the Okoboji region, becomes even more urgent in the face of unrelenting environmental stress. As we have seen, much of this stress is self-inflicted and state supported. We can use our recorded natural history to assess the magnitude of our current environmental problems. Furthermore, we can use this recorded history to set broad goals for restoration efforts.

If we are wise, the future of Okoboji amphibians may be comparatively bright. The U.S. Fish and Wildlife Service, the Okoboji Clean Water Alliance, and the Iowa DNR (Wildlife Division) have begun a program of wetland restoration, primarily for waterfowl production. (Am I the only one who finds it strange that Fisheries DNR is destroying existing wetlands while Wildlife DNR is concurrently creating new ones?) Their ambitious target is to reclaim thirty thousand acres of Iowa wetlands and their surrounding uplands by the turn of the century. Half of these wetlands will be in Dickinson and the surrounding counties. (For more information about the successes and failures of wetland restoration projects, including those in Okoboji, see Galatowitsch and van der Valk's *Restoring Prairie Wetlands*.)

It appears to us that when wetlands are managed for waterfowl, they are indirectly managed for amphibians. In Okoboji amphibians respond rapidly to restored wetlands. Tiger salamanders and leopard frogs colonized the McBreen Marsh complex during the first summer it held water. The second summer, McBreen held these species plus chorus frogs and American toads.

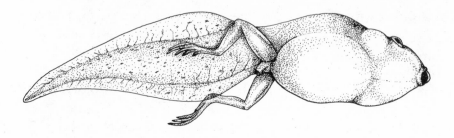

Restored wetlands, if left unmolested, could partially reverse the amphibian declines experienced in Okoboji over the past century. Still, these new wetlands are without exception small, and because of this are likely to go dry during prolonged droughts. Twenty 5-acre wetlands are not the equivalent of one 100-acre wetland, especially during a drought. In Okoboji, all the DNR-impacted wetlands are permanent, while most of the restored wetlands appear to be temporary. Local amphibians will still face the problem of finding drought refugia. We look forward to observing the water regimes of the restored wetlands during the next drought cycle and to following the patterns of amphibian response.

We have visited the natural world of our grandparents and compared it with ours. Pieces are missing. Now, carry this trajectory through to the world of our grandchildren. Do you like what you see? As one of Okoboji's finest ghosts, the spirit of Frank Blanchard is already here. We wish we could say the same thing about his cricket frog. If we work at it, if we care about it, if we learn a wetland ethic and put it into practice, someday we will.

REFERENCES

Anonymous. 1907. Frogs. *OPA Bull.* 2:5.

Anonymous. 1915. News Items. *OPA Bull.* 11:43.

Anonymous. 1922. *OPA Bull.* 18:123.

Anonymous. 1929. *OPA Bull.* 25:37, 116.

Barrett, W. 1964. Frogging in Iowa. *Annals of Iowa* 37:362–365.

Bishop, S. C. 1948. *Handbook of Salamanders: The Salamanders of the United States, of Canada, and of Lower California.* Handbooks of American Natural History Series. Ithaca: Comstock Press.

Blanchard, F. 1923. The amphibians and reptiles of Dickinson County, Iowa. Univ. of Iowa Studies in Natural History. *Lakeside Lab. Studies* 10:19–26.

Bovbjerg, R. V. 1965. Experimental studies on the dispersal of the frog, *Rana pipiens. Proc. Iowa Acad. Sci.* 72:410–418.

Bovbjerg, R. V., and A. M. Bovbjerg. 1964. Summer emigrations of the frog *Rana pipiens* in northwestern Iowa. *Proc. Iowa Acad. Sci.* 71:511–518.

Bradshaw, C. S. 1930. OPA President's Annual Report. *OPA Bull.* 26:37–43.

Brooks, P. 1989. *The House of Life, Rachel Carson at Work.* Boston: Houghton Mifflin.

Brown, M. 1910. Iowa Lakeside Laboratory as a student sees it. *OPA Bull.* 5:11–13.

Carlson, B. M., and D. Stange. 1986. Frogs and waterdogs—The "right-stuff" for big bass, pike and walleyes. *In-Fisherman* 68, Aug.–Sept.:43–62.

Carothers, S. W., and B. T. Brown. 1991. *The Colorado River Through the Grand Canyon, Natural History and Human Change.* Tucson: University of Arizona Press.

Carson, R. 1963. *Silent Spring.* Boston: Houghton Mifflin.

Christiansen, J. L., and R. M. Bailey. 1991. *The Salamanders and Frogs of Iowa.* Nongame Technical Series No. 3. Des Moines: Iowa Department of Natural Resources.

Dickerson, M. C. 1908. *The Frog Book: North American Toads and Frogs with a Study of the Habits and Life Histories of Those of the Northeastern United States.* New York: Doubleday, Page.

Dinsmore, J. J. 1994. *A Country So Full of Game.* Iowa City: University of Iowa Press.

Ehrlich, G. 1986. *The Solace of Open Spaces.* New York: Penguin Books.

————. 1991. *Islands, the Universe, Home.* New York: Penguin Books.

Erickson, D. 1984. Unpublished student project. Iowa Lakeside Laboratory.

Errington, P. L. 1987. *A Question of Values.* Ames: Iowa State University Press.

————. 1987. *Of Men and Marshes.* Ames: Iowa State University Press.

Flader, S. L., and J. B. Callicott, eds. 1991. *The River of the Mother of God and Other Essays by Aldo Leopold.* Madison: University of Wisconsin Press.

Galatowitsch, S. M., and A. G. van der Valk. 1994. *Restoring Prairie Wetlands: An Ecological Approach.* Ames: Iowa State University Press.

Hamilton, H. L. 1939. The biological action of rotenone on lake fauna. *Iowa Acad. Sci.* 44:457–458.

Hammerson, G. A. 1982. Bullfrogs eliminating leopard frogs in Colorado? *Herp. Review* 13:115–116.

Harlan, J. R., and E. B. Speaker. 1987. *Iowa Fish and Fishing.* Des Moines: Iowa Department of Natural Resources.

Harr, D. C., D. M. Roosa, J. C. Prior, and P. J. Lohmann. 1990. *Glacial Landmarks Trail: Iowa's Heritage of Ice.* Des Moines: Iowa Department of Natural Resources.

Hayes, M. P., and M. R. Jennings. 1986. Decline in ranid frog species in western North America: Are bullfrogs (*Rana catesbeiana*) responsible? *Journal of Herpetology* 20:490–509.

Heat-Moon, W. L. 1991. *PrairyErth.* Boston: Houghton Mifflin.

Iowa Sportsman's Atlas. 1994. Compiled and published by Sportsman Atlas Company, Lyton, Iowa.

Kelley, H. M. 1926. The clams of the Okoboji Lakes. *OPA Bull.* 22:53–58.

Kupferberg, S. J. 1993. Bullfrogs (*Rana catesbeiana*) invade a northern California river, a plague or species coexistence? Ecological Society of America's Annual Meeting, Madison. Tempe, Arizona: Ecological Society of America.

Lannoo, M. J., and R. V. Bovbjerg. 1985. Distribution, dispersion, and behavioral ecology of the land snail *Oxyloma retusa* (Succineidae). *Iowa Acad. Sci.* 92:67–69.

Lannoo, M. J., K. Lang, T. Waltz, and G. Phillips. 1993. An altered amphibian assemblage: Dickinson County, Iowa, 70 years after Frank Blanchard's survey. *Amer. Midl. Nat.* 131:311–319.

Larrabee, A. P. 1926. An ecological study of the fishes of the Okoboji region. University of Iowa Studies in Natural History 11 (12):1–35.

————. 1927. The fishes of the Okoboji lakes. *OPA Bull.* 23:112–123.

Lavel, L. B., and A. C. Upton. 1987. *Toxic Chemicals, Health, and the Environment.* Baltimore: Johns Hopkins University Press.

Lean, G., D. Hinrichsen, and A. Markham. 1990. *World Wildlife Fund Atlas of the Environment.* New York: Prentice-Hall Press.

Lendt, D. L. 1989. *Ding: The Life of Jay Norwood Darling.* Ames: Iowa State University Press.

Leopold, A. S. 1986. *A Sand County Almanac.* New York: Ballantine Books.

Lonergan, M. J. 1930. Report on nuisance condition, Upper Gar Lake, Dickinson County. *OPA Bull.* 26:100–106.

Macbride, T. H. 1909. The Okoboji Lakeside Laboratory. *Iowa Acad. Sci.* 16:131–133.

————. 1911. The conservation of our lakes and streams. *OPA Bull.* 6:37–46.

Mac Farland, F. 1911. The Lakeside Laboratory. *OPA Bull.* 6:34–36.

Miller, R. R. 1963. Is our native underwater life worth saving? *National Parks Magazine* 37:5.

Moyle, P. B. 1973. Effects of introduced bullfrogs, *Rana catesbeiana*, on the native frogs of the San Joaquin Valley, California. *Copeia* 1973:18–22.

Oldfield, B., and J. J. Moriarty. 1994. *Amphibians & Reptiles Native to Minnesota.* Minneapolis: University of Minnesota Press.

Pammel, L. H. 1929. Some items from the diary of S. W. Kearny. *OPA Bull.* 25:58–63.

Phillips, K. 1994. *Tracking the Vanishing Frogs: An Ecological Mystery.* New York: St. Martin's Press.

Pister, E. P. 1976. A rationale for the management of nongame fish and wildlife. *Fisheries* 1:1–14.

Ponting, C. 1936. *A Green History of the World.* New York: St. Martin's Press.

Prior, J. C. 1991. *Landforms of Iowa.* Iowa City: University of Iowa Press.

Recher, G. 1992. *Iowa Great Lakes Recreational Map.* Estherville, Iowa: Recher Enterprises.

Schramm, C. W. 1914. Thirty years ago. *OPA Bull.* 10:38–43.

Sharp, A. G. 1908. Untitled [on "Settlement"]. *OPA Bull.* 3:19–20.

Stephens, T. C. 1918. The birds of the lake region. *OPA Bull.* 14:10–17.

———. 1922. Mammals of the lake region of Iowa. *OPA Bull.* 18:47–64.

Stumpel, A. H. P. 1992. Successful reproduction of introduced bullfrogs *Rana catesbeiana* in northwestern Europe: A potential threat to indigenous amphibians. *Biological Conservation* 60:61–62.

Thomas, A. O. 1913. The glacial story of the Lake Okoboji region. *OPA Bull.* 9:17–21.

Tisdale, S. 1994. The only good mongoose is . . . : Rave about Hawaii's wildlife, but please don't say "exotic." *Outside* 19:47–54.

Tweed, S. T. 1938. Unpublished student project. Iowa Lakeside Laboratory.

U.S. Environmental Protection Agency. 1993. *Created and Natural Wetlands for Controlling Nonpoint Source Pollution.* Corvallis: Office of Research and Development.

van der Valk, A. 1989. *Northern Prairie Wetlands.* Ames: Iowa State University Press.

Williams, T. T. 1994. *An Unspoken Hunger: Stories from the Field.* New York: Pantheon Press.

Wright, A. H., and A. A. Wright. 1933. *Handbook of Frogs and Toads: The Frogs and Toads of North America.* Handbooks of American Natural History Series. Ithaca: Comstock Press.

Wylie, R. B. 1912. The aquatic gardens of Lake Okoboji. *OPA Bull.* 7:10–20.

INDEX

SELECTED BUR OAK BOOKS

A Country So Full of Game:
The Story of Wildlife in Iowa
By James J. Dinsmore

Fragile Giants: A Natural History
of the Loess Hills
By Cornelia F. Mutel

Iowa Birdlife
By Gladys Black

Landforms of Iowa
By Jean C. Prior

Land of the Fragile Giants:
Landscapes, Environments,
and Peoples of the Loess Hills
Edited by Cornelia F. Mutel and
Mary Swander

Okoboji Wetlands: A Lesson
in Natural History
By Michael J. Lannoo

Prairies, Forests, and Wetlands:
The Restoration of Natural
Landscape Communities in Iowa
By Janette R. Thompson

Restoring the Tallgrass Prairie:
An Illustrated Manual for Iowa
and the Upper Midwest
By Shirley Shirley

The Vascular Plants of Iowa:
An Annotated Checklist
and Natural History
By Lawrence J. Eilers and
Dean M. Roosa